Heat Storage Systems
for Buildings

Heat Storage Systems for Buildings

Ibrahim Dincer

Ontario Tech. University, Canada

Dogan Erdemir

Erciyes University, Turkey

ELSEVIER

Elsevier
Radarweg 29, PO Box 211, 1000 AE Amsterdam, Netherlands
The Boulevard, Langford Lane, Kidlington, Oxford OX5 1GB, United Kingdom
50 Hampshire Street, 5th Floor, Cambridge, MA 02139, United States

Notices
Knowledge and best practice in this field are constantly changing. As new research and experience broaden our understanding, changes in research methods, professional practices, or medical treatment may become necessary.

Practitioners and researchers must always rely on their own experience and knowledge in evaluating and using any information, methods, compounds, or experiments described herein. In using such information or methods they should be mindful of their own safety and the safety of others, including parties for whom they have a professional responsibility.

To the fullest extent of the law, neither the Publisher nor the authors, contributors, or editors, assume any liability for any injury and/or damage to persons or property as a matter of products liability, negligence or otherwise, or from any use or operation of any methods, products, instructions, or ideas contained in the material herein.

Library of Congress Cataloging-in-Publication Data
A catalog record for this book is available from the Library of Congress

British Library Cataloguing-in-Publication Data
A catalogue record for this book is available from the British Library

ISBN: 978-0-12-823572-0

For information on all Elsevier publications visit our website at
https://www.elsevier.com/books-and-journals

Publisher: Oliver Walter
Acquisitions Editor: Ruth Rhodes
Editorial Project Manager: Barbara Makinster
Production Project Manager: Kiruthika Govindaraju
Cover Designer: Alan Studholme

Typeset by TNQ Technologies

Working together
to grow libraries in
developing countries

www.elsevier.com • www.bookaid.org

Contents

Preface

At present, we are facing great challenges related to energy, economy and the environment that require immediate attention and right solutions. Such solutions are not easy due to their multidimensional and multidisciplinary nature. In this regard, it is required to consider every sector carefully for energy solutions, including buildings. One needs to know that buildings are recognized as one of the critical sectors as they are responsible for about 36% of the energy consumed in the world. Such a high energy demand in return results in some high amounts of greenhouse emissions and also appears to be responsible for about 39% of carbon emissions. Buildings further play a key role to raise people's living standards and comfort levels, as residential and commercial buildings are the places where people spend most of their time. Almost half of the consumed energy in buildings is particularly used for heating, cooling and fresh air supply purposes. Therefore, heat storage systems offer a great potential to manage energy demand and supply, and have been a unique option for buildings. This critical theme is really the hearth of the book, which will focus on the heat storage applications in buildings and take them beyond the conventional systems and applications used in buildings.

This particular book on heat storage systems for buildings will benefit the students, researchers, scientists, and practicing engineers for better understanding the thermodynamic fundamentals, basic concepts about energy sources, energy systems and useful energy outputs, resource utilization criteria, performance assessment and evaluation, system flexibility, economic aspects, environmental dimensions, and sustainable development. It also provides descriptions, models, analyses, and assessments to help better comprehend the topic. Finally, such a rich and diverse content makes it a distinguished book for senior undergraduate students, postgraduate students, researchers, scientists, and practicing engineers in the field.

This book consists of seven chapters starting from energy and its environmental impacts, need for energy storage, energy storage methods, and history of the energy storage. The introductory chapter presents various fundamental aspects and concepts about energy and its use as well as environmental impacts, and it also discusses general energy storage methods briefly. Chapter 2 offers key details of the heat storage techniques along with some fundamental information about energy, heat and temperature. The advantages of heat storage systems are also presented through some illustrative practical applications. Chapter 3 focuses on energy consumption profiles in residential and commercial buildings to emphasize the potential needs for heat storage systems. Chapter 4 defines the methods used for analyzing and modeling the heat storage systems. The fundamentals of thermodynamics, heat transfer, and computational fluid dynamics are presented to the readers. Chapter 5 presents various heat storage systems using different energy sources, which are firstly evaluated through the fundamentals of thermodynamics based on energy and exergy calculations. Then, some key applications of heat storage systems are introduced. This is followed by the computational fluid dynamics applications.

Chapter 6 focuses on heat storage—related applications of artificial intelligence where some artificial neural network methods are introduced. The need for smart energy systems is additionally emphasized. Chapter 7 closes the book with some important conclusions and possible future directions of heat storage systems.

We hope this book provides fundamental information as well as presents new dimensions about heat storage systems and their applications in buildings, and helps the communities implement better practices for achieving better sustainable development targets.

Ibrahim Dincer
Dogan Erdemir

Fundamentals and Concepts

1.1 Introduction

It has been crystal clear to everyone that life is not possible without energy, what really makes a key driver for anything and everything. It has really been an ultimate need for humankind and their economic activities in many sectors throughout the history, more importantly after the industrial revolution. Energy has dominated many things, such as relationships, business and political deals, peace, wars, and terrorist activities, etc., is dominating more at present and will sure dominate far more in the future. Our present civilization has an ultimate responsibility to deal with this matter diligently and fairly.

The increasing world population and the rising living standards and comfort level in almost all sectors have been the reasons behind the globally ever-growing energy consumption. Fig. 1.1 illustrates the total amount of energy consumption around the world from 1965 to 2018. It is clear from Fig. 1.1 that energy consumption in the

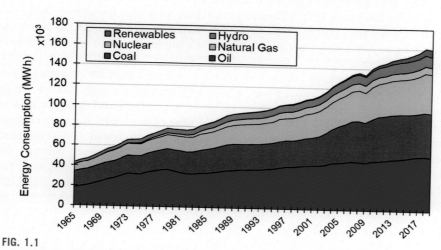

FIG. 1.1

Total energy consumption of the world.

Data from Ref. [1].

Heat Storage Systems for Buildings. https://doi.org/10.1016/B978-0-12-823572-0.00007-2

world is ever increasing substantially every year, and fossil-based fuels meet almost two-thirds of energy demands. Also, the decreases in energy consumption in 1972−75, 1979−83, and 2008 of years have been occurred due to economic recession. From 1965 to 2018, energy consumption has increased approximately four times.

Fossil fuels that meet almost two-thirds of energy demand have some risks such as their limited and nonhomogeneous reserves, environmental impacts, and energy security concerns. These risk factors have forced people toward alternative energy sources and systems such as renewables. Also, with the oil crisis that emerged in October 1973, the interest in alternative energy sources has increased significantly. Especially, solar domestic hot water systems and other solar space heating systems have become prevalent after the oil crisis. The use of renewable energy sources has increased significantly over the past decade. Despite the magnificent advantages of renewables in terms of energy security and environmental impact, the most critical obstacle to their use is that they are noncontinuous energy sources due to their nature. Energy storage (ES) systems have a significant potential to solve the mismatch problem in energy supply and demands periods for energy sources, especially for renewables.

In addition to increasing energy use, the second most important issue in energy consumption is the fluctuating load distribution of energy consumption. Fig. 1.2 demonstrates the electricity demand profile of the United Kingdom (Great Britain) as daily, week, and yearly. As can be seen from Fig. 1.2, the electricity demand of the United Kingdom has fluctuant load distribution during daily, weekly, and yearly periods. On a daily basis (Fig. 1.2a), while the minimum demand is approximately 80% lower than the maximum peak load, the average demand is about 30% lower than the maximum peak load. Similar fluctuations have been observed in a monthly (Fig. 1.2b) and annual (Fig. 1.2c) energy demand profile. The fluctuating energy demand trends are also similar in all developed and developing countries. As seen from Fig. 1.2, the peak energy demands are seen for a limited period of time, but varying drastically.

Extra power plants or energy imports are required to meet the peak energy demands. Extra power plants and energy imports bring with it a high cost for energy supply. The different tariff structures, called triple tariff or multi tariff, are used to cover a part of this high-cost energy. Also, the time-of-use tariffs aim to promote to reduce energy consumption for savings during peak periods and to shift the peak demand from peak hours to off-peak hours. Here, the difference between the maximum, average, and minimum energy demands shows the potential of the ES techniques. ES techniques can be used for balancing energy supply and demand. Thus, the extra power plant and energy import requirements can be decreased.

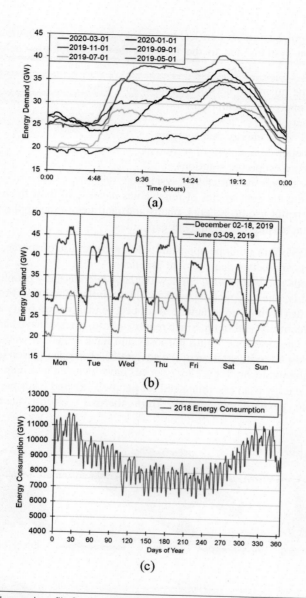

FIG. 1.2

The electricity demand profile for the United Kingdom: (a) daily, (b) weekly, and (c) yearly.

Data from Ref. [2].

ES is recognized as a very critical technology for energy production, conversion, and utilization systems due to the fact that these technologies help achieve the following:

- to offset the mismatch between demand and supply,
- to reduce energy consumption,
- to save energy,
- to provide better energy management,
- to allow better deployment of resources,
- to offer more economic solutions,
- to provide more efficient energy systems,
- to overcome the deficiencies of renewables because of their fluctuating nature,
- to permit better use of systems, plants and facilities in the applications, and
- to avoid increasing the equipment capacities.

In brief, ES offers a unique opportunity to the better use of sources, systems, and useful outputs and better management practices.

ES methods are, by the way, not new but have been used by people since ancient times. Probably, one may recognize ES as old as civilization itself. From past to present, energy has been an important need for people in any form of it such as thermal, mechanic, electricity, and so on. Since recorded times, people have harvested ice and stored it for later use for cooling purposes and have collected woods for the later heating purposes. An illustration of the ancient ES methods is given in Fig. 1.3. Parallel to developing technology, energy needs have changed and increased considerably. Today, energy is an essential need for people's life and takes part in everything from mobile devices to vehicles, from thermal comfort devices to space

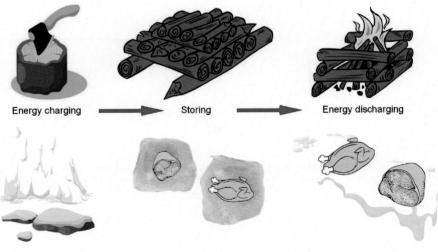

Energy charging ➡ Storing ➡ Energy discharging

FIG. 1.3

An illustration of the energy storage methods in ancient times.

technologies. Increased and changing energy use forced people to use energy efficiently. Thus, multigeneration systems, which generate more than one useful output, have appeared. In addition, discontinuous energy sources, such as renewables, and the differences between the energy supply and demand periods forced people to seek alternative methods to manage the energy supply and demand. Herein, ES methods play a critical role to solve problems in energy use, to benefit efficiently from the energy source, and to recover the lost energy.

More than half of the consumed energy in residential buildings around the world is used for heating and cooling purposes. While many energy sources such as natural gas, electricity, coal, and so on can be used in heating systems, electricity is the main energy source for cooling systems. Also, HVAC (heating, ventilating, and air conditioning) systems consume a substantial amount of energy, and they cause significant operating costs. Besides, they have a great impact on the grid load. ES systems have great potential for reducing operating costs and shifting peak loads.

Although, theoretically, energy can be stored in any energy type, when recent ES storage technologies and practical applications are considered, there are six types of ES which are heat (thermal), mechanical, electrochemical, chemical, biological, and magnetic. The selection of the ES type is conducted based on many criteria such as the amount to be stored energy, storage medium, energy source, storage period, location, physical conditions, environmental conditions, and so on. The heat storage is one of the most common and advanced methods since almost half of the produced and consumed energy around the world is the heat. Also, almost all energy losses in energy systems occur in the form of heat. Heat storage is one of the best ways to recover the heat losses. Especially in heating and cooling systems, it seems to be a significant solution to solve the mismatch between the supply and demand of energy.

Energy has always been an essential need for people, and the amount of energy consumption is increased day by day in the world due to the increasing population, the rising living standards, and the comfort level in all sectors. In parallel to increasing energy consumption, the consumption of fossil fuels increases. Fossil fuel consumption is assumed as one of the primary contributors to global warming. The increasing energy consumption and environmental threats have forced people to use energy more efficiently and to benefit from alternative energy sources. ES methods play a significant role to benefit from energy sources in an optimum way and to provide stable use of renewables stable. We begin this chapter with a summary of the energy use and its environmental impact. Next, the need for ES systems and the importance of ES systems are discussed. Then, ES methods and the advantages of ES are given. Finally, the history of ES is presented.

1.2 Energy Use and Environmental Impacts

Energy has always been a significant and primary need for people since energy systems increase the quality and comfort of life. The aim of the use of energy for people has become to do more work the same duration time or to do the same work in a short time. Therefore, they have benefited from many energy sources from past to present, and only people's perception of energy has changed over time. In ancient times, animal power was an important requirement and social power, and traditional biofuels (wood, animal dung, etc.) are essential fuels. While food and heating had been considered as the primary energy need in the past, transportation had become an essential need with an increasing need to reach more food and heating material. Briefly, while the need for energy is increasing by time, only people's perception of energy has changed over time. Horses and steers have become an essential energy source for transportation and agricultural purposes. With the first industrial revolution and the invention of the steam-powered engine, the concept of energy has become different for people. The term of "horsepower" has come into people's life. The term has been adopted by James Watt to compare the output of steam engines with the power of horses. Horsepower (HP) is a unit of power, and it is used for the power output of engines. There are many different standards and definitions of horsepower. Two common definitions being used today are the imperial horsepower, which is about 745.7 W, and the metric horsepower, which is approximately 735.5 W. With the invention of steam-powered machines, coal played a leading role in energy needs. Oil was the main energy need with the invention of internal combustion engines. Then, natural gas and nuclear power came into people's lives due to the increase in electricity use. Today, electricity, heat, and transportation are significant energy needs for people. These needs meet with the many energy sources such as fossil-based fuels, renewables, nuclear power, etc.

Fig. 1.4 demonstrates the massive increase in energy consumption around the world and added energy sources with time. Energy consumption in the world

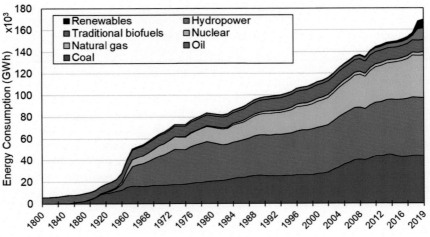

FIG. 1.4

World energy consumption by fuel source, based on Vaclav Smil estimates from energy transitions: history, requirements, and together with BP statistical data for 1965 and subsequent.

Data from Refs. [1–3].

has increased substantially with the industrial revolution. Energy consumption has increased approximately 500 times since the 1800s. While the primary fuel has been traditional biofuels up to the industrial revolution, coal has been one of the main energy sources after the industrial revolution. Oil and natural gas have been added to the type of energy sources with an increasing amount of energy consumption and the use of internal combustion engines. Nuclear and hydropower have been the first alternative energy sources. Although renewable has been used since old times, the amount of consumed renewable energy has not been a significant amount. In the 1990s, renewable energy sources have been involved with a substantial amount.

Fig. 1.5 shows the energy consumptions in the world by sectors between 1990 and 2017. As mentioned in Fig. 1.5, the energy consumption around the world has been increased in all sectors day by day. Especially, the increase in energy consumption in the industry, residential, commercial, and public services sectors is remarkable. When the data for 2017 are considered, while the energy consumption in the residential sector is approximately 30% of all energy consumption, the industry sector covers 40% of all energy consumption. Also, commercial and public services and transportation sectors cover 20% of all energy consumption. A major part of consumed energy in the residential, commercial, and public service sectors is used for heating and cooling purposes. Additionally, most of the consumed and generated energy around the world is heat, and all losses occur as heat. These findings clearly show that heat storage systems have the essential potential to recover losses and to increase system efficiency.

The increase in the variety and quantity of energy needs has brought with it some important problems. One of the most important problems caused by energy use is environmental impacts. Energy supply and demand are related not only to problems such as global warming but also to environmental concerns such as air pollution, ozone depletion, forest destruction, and emission of radioactive substances. Environmental issues should be considered for future people's life and health. Environmental issues should be minimized by better and sustainable systems. There is a close relationship between energy, environment, and sustainable development. ES techniques have a big potential to decrease the environmental impact of energy conversion systems.

During the past few decades, the risks and reality of environmental degradation have become recognizable. The environmental impacts of human activities have raised substantially due to a combination of numerous factors, such as increasing world population, energy consumptions, industrial activities, and so on. These have resulted in various crucial environmental issues, and some of these major environmental problems may be classified as follows [5]:

- acid rain,
- stratospheric ozone depletion,
- global warming and/or climate change,
- air pollution,
- water pollution,
- soil pollution,
- decreased ambient air quality, and
- increased radiation and radioactivity,

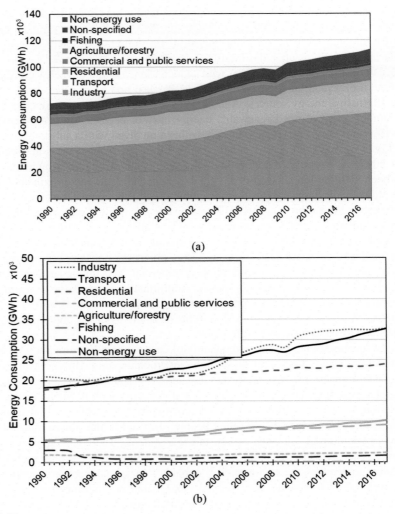

FIG. 1.5

Total final energy consumption by sectors in the world: (a) cumulative sum of energy consumption and (b) the change of yearly energy consumption.

Data from Ref. [4].

which have affected human health and human welfare significantly.

One may note that SO_2, NO_x, CO, CO_2, and volatile organic compound (VOC) emissions are the main indicators for environmental issues. Fig. 1.6 shows the CO_2 emissions according to energy sources and sectors. Among the energy sources, coal is the source that causes the highest carbon emission. 40% of total CO_2 emissions has been occurred by coal-powered energy systems. Oil is the second-highest cause

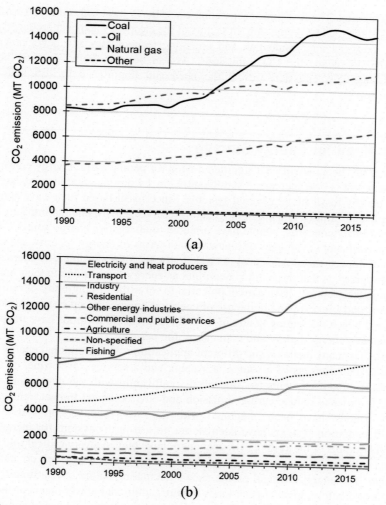

FIG. 1.6

The CO_2 emissions (a) by energy sources and (b) by sectors.

Data from Ref. [4].

of CO_2, and it has covered 37% of total CO_2 emissions. Natural gas has caused the lowest CO_2 emissions among common fossil fuels. 20% of total CO_2 emissions have been released by natural gas–powered systems. Considering the CO_2 emission on a sectoral basis (Fig. 1.6b), the highest CO_2 emission is caused by electricity and heat generation systems. The second-highest emission has been caused by the transport sector, whereas the third-highest emission has been caused by the industry sector. Electricity and heat generation systems have been responsible for almost 40% of total

CO_2 emission. The first-three sectors that cause the highest CO_2 emission cover two-thirds of the total CO_2 emission. In sectors of the electricity and heat producers, transport, industrial and residential, and emerging technologies and developments are required to reduce the CO_2 emissions. As can be seen from Fig. 1.6, the ever-increasing CO_2 emissions reveal the need and importance for ES and energy efficiency.

1.3 Need for Energy Storage

Energy has been/is/will be a significant need for people's life. Energy consumption in the world is ever increasing significantly day by day. Energy demands are generally met by fossil fuel—powered systems. Since fossil-based fuels have some risks such as their limited and nonhomogeneous reserves, environmental impacts, and energy security concerns, people have oriented to alternative energy sources. Despite the magnificent advantages of renewables in terms of energy security and environmental impact, the most critical obstacle to the use of renewables, as alternative energy sources, is that these are noncontinuous energy sources due to their nature. ES techniques primarily have a great potential to solve the discontinuity of renewables. The second-biggest issue in the use of energy is an unbalanced energy demand profile and hence fluctuating nature of it. In this regard, ES methods have the potential to balance energy supply and demand profile periods.

ES has many methods and different applications. A conventional energy conversion system consists of a source, a system, and a service. Over time, each of the aforementioned elements has improved significantly with the developing technology. In order to reduce carbon emission and environmental impact, renewable energy sources have been integrated with the energy conversion systems. Also, multigeneration systems have been started to use in the system for increasing the useful outputs of the system. These changes have played a very critical role in better sustainability. However, energy production and demand profiles do not always match. The energy demand profile shows fluctuating nature over a time frame. As expressed before, to meet peak loads requires additional power plants or the energy imports, and so it requires a high cost. Also, the ever-increasing use of wind turbines and their increasing contribution to nighttime electricity generation will bring about a rising nighttime surplus of electricity generation. This causes waste energy, which is produced by wind turbines. Nuclear power plants also have a similar situation. ES has a high impact on recovering waste energy. Fig. 1.7 demonstrates a novel aspect of energy conversion systems. Integrating ES to the conventional energy conversion system, between the source and system and between the system and service, reduces the waste energy and solves the mismatch between the energy production and demand periods. This rule is called as 3S + 2S = 5S. The rule of 3S + 2S presents a higher efficiency and better sustainability.

ES can in simple terms be defined as "to store energy in a storage medium for a later use." The aim of later use of the energy is to overcome numerous challenges as follows:

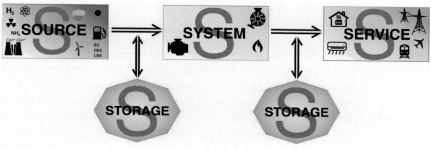

FIG. 1.7

The rule of 5S (5S = 3S + 2S).

Modified from Ref. [6].

- to balance energy supply and demand,
- to decrease the energy consumption and related costs,
- to avoid the peak energy loads,
- to shift the peak energy loads to off-peak periods,
- to reduce the size and capacity of the system equipment,
- to minimize energy losses and benefit from them accordingly if any, and
- to use as backup solution during emergencies.

ES may achieve one of them, most of them, or all of them. Each aim is a significant case for energy efficiency and management applications. This achievement is depending on ES utilization strategy, also called ES scenario. ES utilization strategies will be discussed in the further section. Solar domestic hot water systems may be given as an example of using energy when the energy source is not active. In a solar domestic hot water system, solar energy is stored as heat in the water-filled hot water tank for night usage. Thus, it is possible to benefit from solar energy when the sun is down. This storage can be performed for short term (night, daily, etc.) and for long term (seasonal, yearly, etc.). For instance, solar energy can be charged to rock bed by converting heat during the summer, and the stored energy can be used for heating purposes during the winter. Renewable energy sources are not continuous supplies of energy because of their nature. They need to be stored to continue the use of them. Therefore, ES methods play an essential role to expand the use of renewables.

Electricity is generally supplied with different tariff structures according to the time-of-use period to users. While the electricity unit price is high during peak periods when electricity usage is intense, electricity is provided at a lower price during night hours when electricity usage is low. The aim of this time of use period is to promote energy conservation and reduce peak demand for electricity. ES techniques can be used for reducing energy consumption costs. For example, in cold heat storage systems, energy is charged to a cold medium by using chillers during the night hours when the electricity unit price is lowest. Then, building cooling needs in the expensive electricity period is carried out using stored energy. In this way, the cooling needs of the building can be met in a cheaper way.

To meet peak energy loads is a significant issue due to requirement of high-capacity electricity production and service. Also, peak loads are seen in a very limited period when the daily electricity usage period considered. This limited time and high-capacity demand is quite high according to the average load. Therefore, the power plant or energy imports are needed to meet the peak demands. Fig. 1.2 illustrates the electricity demand profile of the United Kingdom (Great Britain) as daily, monthly, and yearly. As can be seen from Fig. 1.2, on a daily basis, while the minimum demand is approximately 80% lower than the maximum peak load, the average demand is around 30% lower than the maximum peak load. Similarly, on a monthly basis for December and a yearly basis, the difference between the average and maximum demands is similar. Additionally, the difference between the minimum energy demand and the maximum energy demand is almost the same for all countries. Governments should build extra power plants or import energy to meet peak energy demands. Hence, to meet peak loads brings a high cost for the energy supply. The different tariff structures are used to cover this high-cost energy during peak hours. Also, these time-of-use tariffs are used to promote conservation and reduce peak demand for electricity. The difference between the maximum and average energy demands and time-of-use tariffs shows the potential of ES. Shifting the peak energy demands is possible with the ES system. For example, via a pumped ES system, the water is pumped to a higher level during the off-peak electricity hours, and then this water is used for producing energy during peak hours. As an example of heat storage, the cold thermal ES systems such as ice thermal ES systems are used for shifting peak cooling loads to off-peak hours. Thus, the electricity consumed by air conditioners can be moved to off-peak hours, and the effect of cooling needs on the grid loads can be decreased.

ES systems also have the capability to reduce the sizes and capacities of devices used in the energy conversions systems. This benefit is achieved by shifting the peak loads of energy. For instance, heating and cooling systems for a building are designed according to extreme design conditions. These conditions are seen in a short time when the system use is considered. The need for short-term and high peak load requires considerable investment to meet the heating and cooling needs of buildings. With the heat storage methods, the peak heating and cooling loads can be shifted to off-peak hours by using proper heat storage strategy. Thus, both the capacities and sizes of the devices and the energy cost can be reduced by the ES methods.

ES systems can regain the lost energy from nonstop power plants such as wind and nuclear energy. During the night hours, energy consumption decreases since energy use in sectors industrial, residential, public, transportation, and so on are decimated. Nevertheless, the nonstop power plants, such as wind and nuclear energy, continue energy production. The produced energy by these power plants is not used because of lower demand, and the energy is wasted. Thanks to the ES methods, this waste energy can be recovered. The pumped ES, compressed air ES, and ice thermal ES systems can be given as an example for this purpose. In the future, the ever-increasing use of wind turbines and their increasing contribution to nighttime electricity generation will bring about an increasing nighttime surplus of electricity generation, which adds to the attraction of shifting power consumption to night hours. This situation will increase interest in the time-shifting applications in energy use such as electrical ES in batteries and cold storage in ice, also called ice storage.

   Lastly, ES techniques may be used for coming through emergencies and reducing the cost and size of devices that are used in emergency situations. For instance, in buildings where cooling is important and critical, such as data processing centers, supercomputer centers, stem cell and embryo centers, and museums, there are one or more substitute devices for cooling systems and their generators. This brings along with it a substantial initial investment and service costs. Ice storage or other cold heat storage techniques can be used for emergencies. Thus, both cooling costs and the costs of devices used for emergencies can be reduced.

   ES techniques are also significant for smart energy portfolio highlighted in Dincer [7]. A smart energy portfolio consists of nine key options/solutions, which are storagization, intelligization, exergization, greenization, renewabilization, hydrogenization, integration, and multigeneration [7]. The aim of the smart energy portfolio is to manage smartly the entire energy spectrum, which are energy fundamentals and concepts, energy materials, energy production, energy conversion, and energy management. Each branch is significant to achieve a more sustainable future, and all branches should be added to energy systems. Here, the storagization plays a critical role in offsetting the mismatch between demand and supply and to operate the systems in a more efficient, economic, and environmentally sound manner. Also, the intelligization is a critical requirement for optimal management of systems. As mention in Fig. 1.8, the intelligization is required to manage the 5S systems, which consists of a source, system, service, and storages. The intelligization can provide real-time management considering the environmental, design,

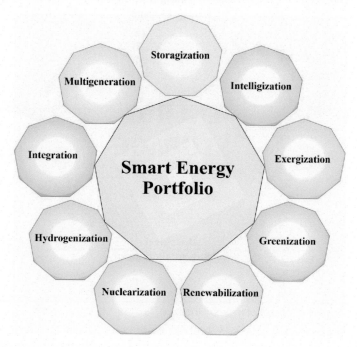

FIG. 1.8

Nine branches of smart energy portfolio.

Modified from Ref. [6].

and operating conditions to maximize or minimize a set of purpose functions, such as efficiency, operating cost, initial cost, and so on.

1.4 Energy Storage Methods

ES can basically be defined as "to store energy in a storage medium for later use." Theoretically, it is possible to store energy in each form of energy. When practical applications and storing capacities are taken into consideration, ES can be classified into six branches, as illustrated in Fig. 1.9.

Note that each of these ES methods has some important advantages and also disadvantages depending on their area of usage and each ES method has many and different subapplications. The classification of ES methods according to their energy formations and storage materials is shown in Fig. 1.10. Many ES techniques are still under improvement. We shall discuss them by categories, grouping together those techniques that store energy in the following forms: mechanical, heat, chemical, electrochemical, magnetic and electromagnetic, and biologic, as shown in Fig. 1.10. In this chapter, all ES methods will be introduced briefly except for

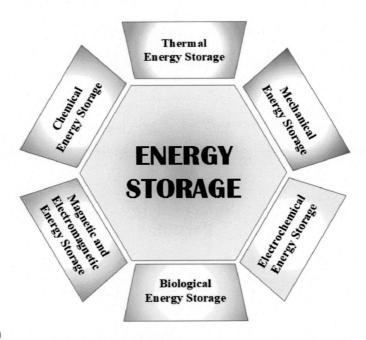

FIG. 1.9

The classification of energy storage methods.

heat storage. Heat storage and its applications will be given in detail in the next chapter.

1.4.1 Mechanical Energy Storage

Mechanical energy can be stored in the form of mechanical energy types which are kinetic and potential energies. Mechanical energy can be stored as kinetic energy by changing the linear and rotational velocities of an object. Also, it can be stored as potential energy by elevating an object or compressing a spring, gas, etc. In practical relevance, it is impossible to store a large amount of energy in linear motion due to requirement of long distance or large mass. However, it is quite possible to store energy by using rotational velocity of an object in the form of kinetic energy. When the practical applications are taken into consideration, mechanical ES techniques can be classified as follows:

- Pumped hydro ES
- Compressed air ES
- Flywheels

Although Potter's wheel and springs of tower clocks that are developed several thousand years ago have been the first form of the mechanical ES, they are not accepted as a form of mechanical ES due to their ES capacities. In addition to this, coil springs used in toys and clocks are basic examples of mechanical ES, too. In this book, under mechanical ES techniques, these three methods, namely pumped ES, compressed-air ES, and flywheels, are discussed.

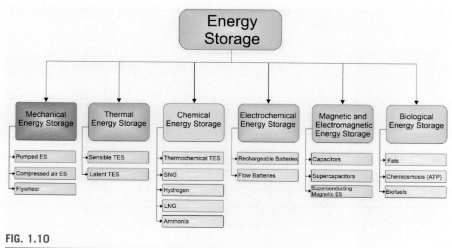

FIG. 1.10

The classification of energy storage methods.

1.4.1.1 Pumped hydro energy storage (hydrostorage)

It is important to note that pumped hydro ES, also called as hydrostorage and pumped storage, is a simple ES method technically. The main purpose of the pumped ES is to balance energy supply and demand by shifting the peak loads of off-peak hours. The schematic view of the pumped ES is shown in Fig. 1.11. The working principles of pumped hydro ES are the following:

- **Charging period:** During the off-peak hours, generally at night, when the energy demands are low and the unit cost of electricity is cheaper, pumps are operated to pump the water from a lower level to the upper level.
- **Storing period:** The elevated water is kept in the upper reservoir until the peak of the energy.
- **Discharging period:** During peak hours, when the energy demands are high and unit cost of electricity is higher, the pumped water in the upper level flows to the lower level, and this flow turns the turbine. A generator assembled to the turbine is used for generating electricity.

Thus, peak electricity loads can be shifted from peak hours to off-peak hours. The difference between energy peak demand and off-peak period can be reduced. In this way, it is possible to reduce the additional power plant requirements and the amount of imported energy.

Pumped hydro ES may also be used for storing renewables. For this purpose, a solar-powered pump can be used for pumping the water from a lower reservoir to the upper reservoir. At night, when there is no solar energy, the pumped water turns the turbine to generate electricity. It is a simple way to store solar energy for the use of electricity. Similarly, wind-powered pumps can be used for pumping the water to the upper reservoir. And the pumped water can be used to balance electricity production due to varying wind speed.

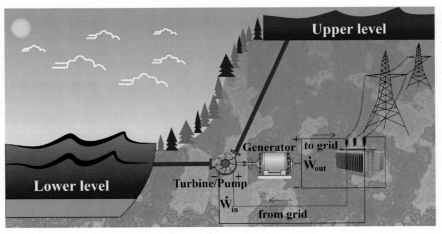

FIG. 1.11

The schematic view of the pumped hydro storage.

In the pumped hydro ES, while the lower-level reservoir can be a lake, river, dam, or sea, the upper-level reservoir can be either an artificial dam, lake, etc. or a natural lake, river etc. The biggest disadvantage of the pumped hydro storage system is the availability of a suitable place for the facility. The fact that lower- and upper-level reservoirs are natural structures can significantly reduce the installation cost of the system. The advantages of the pumped hydro ES can be listed as follows:

- The principles of the system are easy.
- It consists of simple equipment.
- Energy can be generated from stored water (energy) in a few seconds with full capacity.

Underground water reservoirs can be used in pumped hydro ES. Underground water reservoirs are a natural lower-level reservoir. When water is pumped to ground-level artificial or natural reservoirs, energy can easily be stored thanks to distance between the underground and ground levels. The schematic view of the underground pumped hydro ES is shown in Fig. 1.12. The working principles of the underground pumped hydro ES are similar to the pumped hydro ES system. The upper surface reservoir can be an existing body of water or an artificial lake formed by dikes and dams. The lower reservoir is a large cavern excavated in hard solid rock. Underground systems can be more acceptable since the environmental impact is decreased. The area in underground pumped hydro storage needing the greatest development and having the greatest cost is the lower reservoir [8].

In the past, there were about 40 pumped hydro ES power plants, which have a high capacity, around the world like in Bath County, the United States (2710 MW), and Kanagawa, Japan (2700 MW). Today, there are approximately 500 pumped hydro ES power plant around the world, and their capacities vary

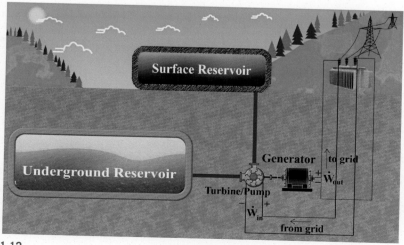

FIG. 1.12

The schematic view of the underground pumped hydro storage.

from a few MW to a few thousand MW. When the total capacity of the ES systems around the world is considered, pumped hydro ES covers 97% of the global capacity of ES power and 99% of the stored energy around the world [9]. According to 2018 data, China has the highest installed capacity of pumped hydro ES in the world with 29.9 GW [10]. If geographic infrastructure and water resources are available, pumped hydro ES is a convenient method to lower the peak of electricity or store renewable energy sources.

The maximum power ($\dot{W}_{max}$) to be generated from a pumped hydro ES power plant is calculated as

$$\dot{W}_{max} = \rho\, \dot{V}_{down}\, g\, H \tag{1.1}$$

Here, ρ is density of the water, $\dot{V}$ is the downward flowrate of the water, g is the gravity, and H is the height difference between lower and upper reservoirs. Eq. (1.1) defines the ideal power capacity of the pumped hydro ES system.

In a pumped hydro ES system, the pump power is the most significant parameter that has a great impact on the efficiency of the system. Pumping power for a pumped hydro ES is calculated as

$$\dot{W}_{P1} = \frac{\rho\, g\, H\, \dot{V}_{up}}{\eta_{P1}} \tag{1.2}$$

where $\dot{V}_{up}$ is the upward flowrate of the water and η_{P1} is the efficiency of the pump.

As expressed before, Eq. (1.1) neglects all irreversibility in the system. In a pumped hydro ES system, there are three main parameters that cause the losses: the losses in hydroturbine, generator, and piping. The net power output of the pumped hydro ES plant can be calculated as

$$\dot{W}_{net\ output} = \dot{W}_{max}\, \eta_{ht}\eta_{gen}\eta_{piping} \tag{1.3}$$

Here, η_{ht}, η_{gen}, and η_{piping} denote the efficiencies of hydroturbine, generator, and piping.

Consequently, energy efficiency of a pumped hydro ES is calculated as

$$\eta_{pumped\ ES} = \frac{\sum_{t=0}^{t_{disch}}\left[\dot{W}_{net\ output}\, t_{disch}\right]}{\sum_{t=0}^{t_{ch}}\left[\dot{W}_{pump}\, t_{ch}\right]} \tag{1.4}$$

Here, t_{ch} and t_{disch} denote the durations of charging and discharging periods. It is clear from Eq. (1.4) that the charging and discharging durations have a significant impact on the system performance. These periods are determined according to the electricity tariff hours. Generally, while the duration of t_{ch} is 10–14 h, t_{disch} is 10–14 h. To reach the higher efficiency in a pumped hydro ES system, t_{ch} should be as maximum as possible.

1.4.1.2 Compressed air energy storage
The working principles of the compressed air storage systems are largely equivalent to the pumped hydro ES system. The aim of the compressed air storage system is to

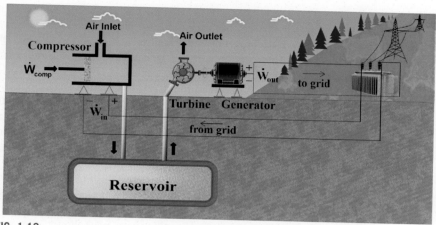

FIG. 1.13

The schematic view of the compressed air storage.

shift the peak energy demands to the off-peak demand period or to store renewables for later use, as same as pumped hydro ES. The schematic view of a compressed air ES system is shown in Fig. 1.13. Basically, a typical compressed air ES system consisted of compressor, turbine, generator, and pressurized reservoir. The working principles of pumped hydro ES are the following:

- **Charging period:** During the off-peak hours, generally at night, when the energy demands are low, and the unit cost of electricity is cheaper, high-capacity compressors are operated to compress the air into a reservoir. This process can also be operated when a renewable energy source is available if the aim of the storage is to store renewables for later usage.
- **Storing period:** The compressed air is kept in the insulated reservoir until energy is needed.
- **Discharging period:** At peak hours when the energy demands are high and the unit cost of electricity is higher or during periods when the renewable energy source is not active, the air is released into the atmosphere by passing through a turbine. A generator assembled to the turbine is used for generating electricity. Also, the pressurized air can also be used in gas turbine systems to reduce the compressor work.

Although the working principle of the compressed air ES seems easy, there are two critical points for the compressed air ES system. The first point is the efficiency of compressors. The reversibility in compressors is quite higher than the pumps. The second critical point is to find a convenient reservoir for storing the compressed air such as natural caverns, old oil or gas wells, porous rock formation, or pressured vessels. The main performance criterion for a compressed air ES is the pressure of the stored air. Higher pressure means higher amount of stored energy. However, air leakage from the reservoir should be considered during the storing period, and it should be minimized. Previous studies indicate that the cost of the system is comparable with that of pumped hydro ES, but the requirement of large caverns or pressured vessels limits the usefulness of this approach to regions where the convenient reservoirs exist. High amounts of energy can be stored by compressing the air in underground caverns.

In a conventional gas turbine, high-pressure hot gas is supplied, and about two-thirds of the total power output is used to drive the compressor. Namely, the compressor is going to use a big portion of the stored energy. A compressed air ES system can reduce the compressor work of a gas turbine, as it already has compressed air. Also, heat recovery can be applied to multistage compressors to increase the useful outputs.

The first compressed air ES plant in the world has been built in 1978 at Huntorf in Hamburg, Germany. In this facility, during the charging period, the air is compressed to around 47.7 kPa in two caverns, which have total storage volume of 283,179 m^3. In the discharging period during electricity peak hours, the air is released and heated by natural gas. Then, it is expanded in high- and low-pressure turbines. The capacity of the system is 290 MW for up to 2 h. The fuel consumption can be reduced by about 25% via the air storage system. As a recent application of compressed air ES, in Ontario, Canada, a startup has built a pilot facility at 1 MW, and they are currently working on some other high-capacity facilities (Hydrostor, 2020 [11]).

A novel compressed air ES method, which stores energy in submerged airbags shown in Fig. 1.14, has been announced by Hydrostor [11]. In this system, the air

FIG. 1.14

Underwater compressed air energy storage system.

Modified from Ref. [11].

has been compressed into airbags submerged. Thus, the compression process can be performed isothermally since the surrounding water of airbags acts as a heat sink. Also, the hydrostatic pressure provides to keep the isobaric condition.

1.4.1.3 Flywheels

Flywheel is one of the oldest ES techniques. Fig. 1.15 demonstrates the schematic view of a typical flywheel used in the practical ES application. A flywheel consists of a rotating mass called the flywheel rotor, magnetic bearing, and motor/generator group. The air inside the flywheel is generally vacuumed to reduce the air friction. The flywheel stores the energy in the form of kinetic energy. The working principles of flywheel ES are the following:

- **Charging period:** The speed of the rotor is increased by the electric motor. The kinetic energy content in the flywheel is increased by increasing angular velocity of the rotor. The charging period is completed when the angular velocity limit is reached.
- **Storing period:** Flywheel is continuing to rotate during the storing period, and the velocity of the flywheel and kinetic energy content reduces. To minimize the energy losses, the magnetic bearing and vacuum environment are used in the flywheel.
- **Discharging period:** When the energy is required, the generator is driven by the flywheel shaft. Thus the stored kinetic energy is converted to electricity.

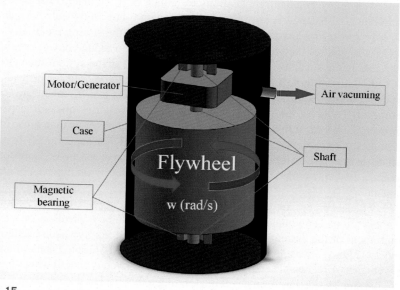

FIG. 1.15

The schematic view of a typical flywheel used for ES purpose. *ES,* energy storage.

Modified from Ref. [12].

The flywheel has a rotating mass, called as flywheel or rotor, generally axisymmetric, and it stores the energy in the form of kinetic energy by changing the angular velocity of the mass. Rotary kinetic energy change in the flywheel is defined as follows:

$$E = \frac{1}{2} I \left(w_1^2 - w_2^2 \right) \tag{1.5}$$

where E is the stored energy in the flywheel, I is the flywheel moment of inertia, and w_1 and w_2 are the initial and final angular velocities for the flywheel, respectively. Whereas the angular velocity increases during the charging period, it decreases during the discharging period. Active magnetic bearings are used for reducing the friction in the shaft bearing. Also, the air inside the flywheel is vacuumed to reduce the air friction. A motor/generator is directly connected to the flywheel rotor to charge and to discharge energy. The bidirectional converter is connected to the motor/generator. The flywheel rotor is a very critical element of a flywheel as it stores energy within the body of its in high angular velocity. The design and strength of the rotor is a significant issue. While steel-based rotors are used in first flywheels, composite-based rotors are used recently to increase the strength of the flywheel.

Although flywheels are one of the oldest ES methods and have not been commonly used in practical applications, today, they have a significant potential to store electricity with the developing technology of its. They also have a big potential to reduce peak energy loads and higher power requirements. Today, flywheels are used in many practical applications to reduce peak demand, to reduce power requirement, to balance energy supply and demand, etc. Flywheels are generally used for short-term electrical storage. The purpose of the storing electricity in flywheels is to control the voltage and ampere changes, to reduce power requirement of the energy conversation system, or to gain the losses in the system such as regenerative braking system. Batteries and ultracapacitors can be used for instead of flywheels. The most advantageous of the flywheels are cost and lifetime. The cost and lifetime of flywheels are quite higher than those of batteries and ultracapacitors. However, the ES capacity and duration for flywheels are very low than those of other techniques. Consequently, when the short-term and higher-capacity ES is needed, such as in the fast-charging station, the use of flywheel as an ES solution may become a convenient method. Flywheels can be used as a single flywheel or a bundle of them.

Flywheels can be integrated into renewable energy applications as a short-term ES application. The voltage and amperage changes can be controlled, or the power requirement of the system can be significantly reduced via flywheels. Erdemir and Dincer [12] have investigated the use of flywheels in fast-charging stations for electric vehicles. A renewable energy-based system integrated with a flywheel-based ES fast-charging station has been developed. Using a flywheel as a short-term ES application has reduced 60% and 72% of the power requirement for the investigated two case studies. In n integrated flywheel system, periods of charging, storing, and discharging play a critical role in the performance of the system. Therefore, they should be arranged carefully.

1.4.2 Heat Storage (Thermal Energy Storage)

Heat storage is to store energy by converting any energy source to heat in a storage medium. Heat ES methods are one of the most mature and advanced ES methods. Since the main purpose of this book is to discuss the heat storage systems and their building application, the heat storage methods are discussed in the next chapter.

1.4.3 Chemical Energy Storage

It is possible to store energy in one or more chemical compounds as a result of one or a series of chemical reaction that absorbs or releases energy. The ES performed in this way is called chemical ES. Chemical energy is stored in the chemical bonds of atoms and molecules, called chemical fuels. The most common chemical fuels are ammonia, hydrogen, liquefied natural gas, and synthetic gases. Thermochemical ES is one of chemical ES, and heat is used for charging and discharging. While it is possible to obtain the heat and electricity by chemical fuels, the heat is the result of a thermochemical ES system. Also, it is possible to generate electricity directly by electron transfer reactions in a chemical ES. The most significant advantage of chemical energy is the capability to store the high amount of energy, in other words, energy density. The common chemical ES methods are discussed in this section.

1.4.3.1 Hydrogen

Hydrogen is a significant material for energy systems due to a clean, nontoxic, and its high energy content. The chemical energy content of hydrogen is 142 kJ/kg, and this value is higher than other hydrocarbon-based fuels. Hydrogen can be obtained with steam reforming and gasification processes from fossil fuels, thermochemical water splitting and high-temperature electrolysis processes from nuclear energy, and electrolysis and gasification processes from renewable energy sources. It releases just water vapor as an emission after combustion reaction [13].

Hydrogen can be stored in two different ways which are forms of compressed gas or liquid and material-based storage. The classification of hydrogen storage methods is demonstrated in Fig. 1.16. High-pressure vessels are required to store hydrogen in the form of compressed gas. Generally, hydrogen is compressed up to 350–700 bars. While the energy density per unit mass increases with increasing storage pressure, the cost of the storage vessel increases. To store hydrogen in the form of liquid phase, cryogenic tanks are required as the boiling point of hydrogen is −252.8°C under atmospheric conditions. Materials-based hydrogen storage is classified as metal hydrides, chemical storage, and sorbent materials. It is possible to store hydrogen in solids by adsorbing to the surface or absorbing to within it. Metal hydride research focuses on improving the energy density, adsorption/

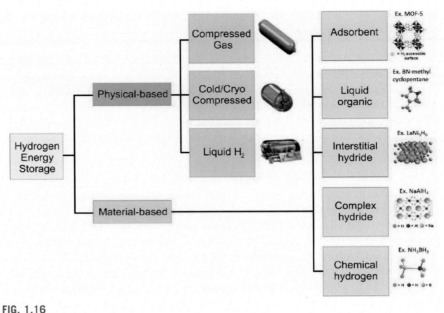

FIG. 1.16

Classification of the hydrogen storage methods.

Modified from Ref. [14].

desorption kinetics, life cycle, and reaction thermodynamics of potential material candidates. Chemical hydrogen storage materials research focuses on enhancing the energy density and transient performance and reducing the release of volatile impurities. Sorbent materials research focuses on increasing effective adsorption temperature and the energy density and optimizing the pore size, pore volume, and surface area of materials.

During the charging period, hydrogen is produced using methods discussed in previous paragraphs by using fossil-based fuels, renewables, nuclear power, etc. Then, the produced hydrogen is stored by using hydrogen storage methods shown in Fig. 1.16 in the storing period. When the energy is needed, the stored hydrogen is burned to generate electricity. A schematic view of the hydrogen ES system is shown in Fig. 1.17. A typical system consists of a hydrogen generator, hydrogen storage tank, and a fuel cell [13].

1.4.3.2 Ammonia

Ammonia (NH_3) as a carbon-free fuel has approximately two times higher energy density per unit volume than liquid hydrogen. In addition to the high energy density, it is a remarkable energy material since it is easier to store, ship, and distribute according to hydrogen. It contains 17.8% of hydrogen. It can be produced from renewable hydrogen and nitrogen separated from the air and can be easily converted to hydrogen and nitrogen. Ammonia can also be used as an alternative

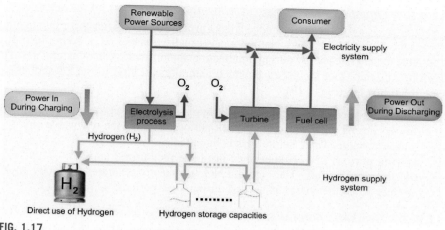

FIG. 1.17

Hydrogen energy storage systems.

Adapted from Ref. [15].

fuel. However, low flammability, high NO_x emission, and low radiation intensity are significant barriers to use. Today, ammonia can be burned in turbines and cofired with pulverized coal and in furnaces. It is also used for obtaining H_2 to use in fuel cells.

For the use of ammonia as an ES application, in the charging period, ammonia is produced by using ammonia production methods. In the storing period, it is kept in a container. When the energy is needed, it is burned or converted to hydrogen and electricity is generated in fuel cells. A solar-driven ammonia-based ES system is shown in Fig. 1.18.

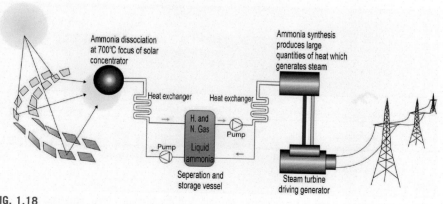

FIG. 1.18

Solar thermochemical ES system using ammonia cycle. *ES*, energy storage.

Modified from Ref. [16].

1.4.3.3 Liquefied natural gas

Natural gas is a significant fuel since it has lower carbon emission according to other fossil-based fuels (Fig. 1.6a). Natural gas is usually transported in liquid form. It is possible to generate energy during liquefaction and regasification of natural gas. The working principles of the use of liquefied natural gas (LNG) for an ES application are the following (Fig. 1.19):

- **Charging period:** Natural gas is compressed by the cryogenic compression process during the electricity off-peak periods or when the renewable energy source is active. Thus, natural gas is liquefied.
- **Storing period:** LNG is kept in an insulated pressurized tank.
- **Discharging period:** The stored LNG is used to generate electricity in gas turbines when the energy is needed during peak periods.

1.4.3.4 Synthetic natural gas

Synthetic natural gas (SNG), also known as substitute natural gas, is one of the most promising technologies in that field. SNG is an alternative which has similar properties of natural gas. Therefore, it plays a critical role in energy systems. SNG means the partial conversion of solid feedstock with gasification followed by gas conditioning. The process of SNG synthesis and gas upgrading is similar to that of natural gas [9]. SNG is produced with a gasification process that is a non-combustion heating process that turns solid carbon fuels into hydrogen, CO_2, and CO. If the hydrogen is created by electrolysis, the process is referred to as power-to-gas or power-to-X, and the resulting product is e-gas or syngas. This process is used for surplus renewables. If the raw material used is plant cellulose, the process is thermochemical SNG production and the resulting gas is called bio-SNG. If

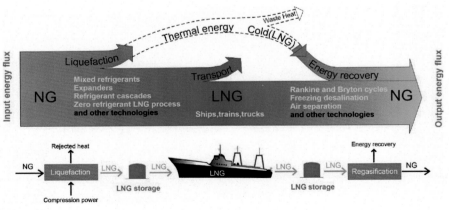

FIG. 1.19

The schematic view of LNG producing, storing, and gasification. *LNG*, liquefied natural gas.

Modified from Ref. [17].

feedstock used is natural anaerobic digestion of organic materials, it is called as biochemical SNG, and the result is the bio-SNG or biogas. SNG is a critical fuel, as it is flexible, storable, and transportable fuel, and it can be produced with renewables. It can be used for storing renewables.

1.4.3.5 Thermochemical energy storage

Thermochemical ES is performed by working endothermic and exothermic chemical reactions in a cycle. The working principles of thermochemical ES are the following:

- **Charging period:** An endothermic chemical reaction is performed by using heat during the off-peak periods or when the renewable energy source is active.
- **Storing period:** The chemical compound obtained in the charging period is kept in a container that provides steady condition for the chemical compound.
- **Discharging period:** The stored chemical is allowed to exothermic reaction and heat is produced. The energy requirement is met with the heat produced.

Thermochemical ES has a significant potential to store energy due to the high energy density and capability of the storage of renewables. However, the most critical disadvantage of thermochemical ES is to keep the unstable chemical compound obtained in charging period during the storing period.

1.4.4 Electrochemical Energy Storage

Electrochemical power sources convert chemical energy to electrical energy and heat. Electrochemical ES is an ES method used to store electricity under a chemical form. This storage technique benefits from the fact that both electrical and chemical energy share the same carrier, the electron. This common point allows limiting the losses due to the conversion from one form to another [18]. In the electrochemical ES, at least two reaction partners undergo a chemical process. The result of this reaction is obtained as electric current and voltage. When practical applications are considered, electrochemical ES has two common applications that are rechargeable batteries and flow batteries. Rechargeable batteries consist of one or more electrochemical cells in series. Flow batteries consist of the ES material that is dissolved in the electrolyte as a liquid. Both are discussed in the following.

1.4.4.1 Rechargeable batteries

Batteries electrochemically store and release the energy as electric energy. Today, batteries play a critical role in energy conversion systems since they are used in a wide range of sectors such as automotive, space, building, renewables, and so on. With the spread of electric and hybrid vehicles, the interest in batteries has increased. It is also one of the important methods to surpass renewable energy sources. In addition, batteries have started to be used in place of building generators. Therefore, many researchers and technology developers have focused on batteries and their packaging processes. There are many types of batteries and the most common ones are zinc—air, nickel—metal hydride, lithium ion, nickel—iron, nickel—zinc, and lead—acid. Their specific energy on mass basis ranges between 172.8 and 828 kJ/kg. Among batteries, in recent years, lithium ion batteries are commonly preferred due to having higher energy density, bad memory effect, lower mass density, and lower self-discharge rates.

Heat generation in the batteries is the most significant issue because increasing temperature reduces the capacity of the lithium batteries. Therefore, the temperature of batteries should be kept under control during the charging and discharging conditions. The capacity fade of battery cells accelerates at higher battery operation temperatures. High operating temperatures can cause overheating, which leads to thermal runaway. High temperature on batteries may cause an explosion or fire. Two main ways of solving the thermal issues of the batteries are to reduce the heat generation rate or to increase the heat dissipation rate. However, the high energy demand and fast charging conditions are considered, and these processes are not possible to achieve. Therefore, batteries should be cooled in an effective way. There are many studies in the open literature that develop battery thermal management systems.

1.4.4.2 Flow batteries

Flow batteries are a type of electrochemical ES, which consists of two chemical components dissolved in liquid separated by a membrane. Charging and discharging of batteries occur by ion transferring from one component to another component through the membrane. The biggest advantages of flow batteries are the capability of pack in large volumes. Interest in flow batteries has increased considerably with increasing storage needs of renewable energy sources. High-capacity flow batteries, which have giant tanks of electrolytes, have capable of storing a large amount of electricity. However, the biggest issue to use flow batteries is the high cost of the materials used in them, such as vanadium. Some recent works show the possibility of the use of flow batteries. The schematic view of the flow battery, an integrated ES system, which is used to store renewable energy, is shown in Fig. 1.20. The flow batteries store electricity in the tanks of liquid electrolyte that is pumped through electrodes to extract the electrons. The flow batteries store electricity in the tanks of liquid electrolyte that is pumped through electrodes to extract the electrons. During the charging period, PV panels, wind turbines, or grid input is used for providing electrons to recharge the electrolyte. The electrolyte is stored in the tank during the storing period. In the discharging period, the liquid electrolyte is pumped through electrodes to extract the electrons and to generate electricity.

1.4.5 Magnetic and Electromagnetic Energy Storage

ES in magnetic and electromagnetic materials can be performed in capacitors, supercapacitors, and superconducting magnetic devices when common ES applications are considered. Magnetic and electromagnetic ES can be classified as capacitors, supercapacitors, and superconducting magnetic ES. They are used for storing electricity.

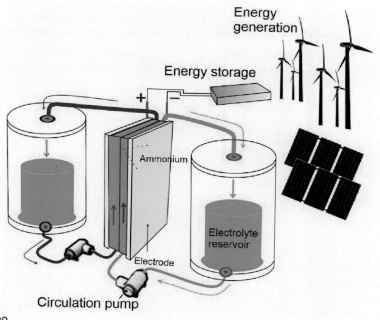

FIG. 1.20

The schematic view of flow batteries.

Adapted from Ref. [19].

1.4.5.1 Capacitors

Capacitors, which consist of two metal plates and a nonconducting separator layer between them, are used for storing electricity. They store electricity on the plate surfaces of metalized plastic film or metal electrodes. Although they have a low energy density, they can supply high electric current for a short time.

1.4.5.2 Supercapacitors

Since the capacitors are not possible to be used in ES systems due to their low energy density, the researchers focused on supercapacitors. According to classical capacitors, the energy density of supercapacitors is quite higher. Supercapacitors are a capacitor that has a double layer without a gap. Being without a gap between the layers is the main structural difference according to classical capacitors and batteries. Supercapacitors have a thin layer of electrolyte and a large surface area—activated carbon structure. The most significant advantage of supercapacitors according to batteries is a high-power output and a high number of charges/discharges.

1.4.5.3 Superconducting magnetic energy storage

Superconducting magnetic ES is a technique to store energy with the magnetic field that is created by the flow of direct current in a superconducting coil. This coil is

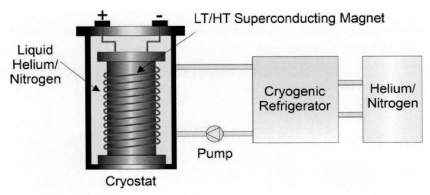

FIG. 1.21

The schematic view of superconducting magnetic ES system. *ES*, energy storage.

Adapted from Ref. [20].

cryogenically cooled below the superconductor critical temperature of the coil material. The schematic view of a superconducting magnetic ES system is demonstrated in Fig. 1.21. A typical superconducting magnetic ES consists of a superconducting coil (cryostat), cryogenic refrigerator, and gas vessel. In the superconducting magnetic ES, when energy charged to the coil, the stored magnetic energy can be stored indefinitely and the stored energy is released by discharging the coil. The form of charged and discharged current should be DC. Therefore, a converter is required to integrate into the system. The average charging/discharging cycle efficiency for a superconducting magnetic ES is about 95%. Although it has high efficiency, the biggest disadvantage of it is the high cost of the superconducting coil and its cryogenic cooling. Therefore, the superconducting magnetic ES method is used for short-term electrical storage.

1.4.6 Biological Energy Storage

Biological ES is significant for livings, from single-cells living organisms to human organisms and from viruses to bacteria. It is necessary for the continuation of life because the energy required to continue the activities of the cells is stored with biological ES. Biological ES can be categorized follows:

- Fats
- Chemiosmosis (adenosine triphosphate [ATP])
- Fuels of biological origin (waxes, oils, biodiesel)

An illustration of the mechanism of biological ES process in a human body is given in Fig. 1.22. In the human body, energy is stored in the form of fat which is the body's most concentrated source of energy. Fatty acids are produced during the digestive and stored in the body. These fatty acids are transported to muscles with blood flow for burning and producing energy. ATP is stored in the muscles

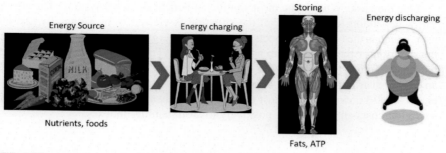

Storing

Energy Source

Energy charging

Energy discharging

Nutrients, foods

Fats, ATP

FIG. 1.22

An illustration of the mechanism of biological ES for a human body. *ES*, energy storage.

for short-term and relatively high-energy demands. Carbohydrates, proteins, and fats are used for producing ATP in the human body. However, carbohydrates and proteins are not stored for producing ATP. Biofuel is produced from biomass that is any organic matter obtained from plants and animals. Biofuels can be produced in gas or liquid form. They are used in almost all energy conversion systems such as from a power plant to vehicles. Since this book focuses on the heat storage and its applications, the listed methods are not discussed in detail.

1.5 Advantageous of Energy Storage

ES techniques provide an optimum way to benefit from energy sources by balancing energy supply and demand, minimizing the losses, extending the energy source active time, lowering capacities of the devices, etc. ES methods have great potentials to enhance energy conversion processes efficiently and manage energy sources sustainably. Some important advantages of ES techniques can be listed as follows:

- Reduces the energy consumption costs
- Increases the efficiency of the system's devices
- Extents the period of use of energy source
- Decreases the capacity and size of the system's equipment
- Reduces the initial investment and service costs
- Balances the energy supply and demands
- Rises generation capacity
- Increases the flexibility of working conditions and hours
- Reduces the fossil-based energy consumption
- Reduces the carbon emission and environmental impacts

An ES system may achieve one, a few or all these benefits depending on the utilization strategy of the ES system. The advantages of ES systems have been discussed with the practical examples in Sections 1.1 and 1.3.

One may ask: What is the most significant benefit of an ES system? The answer varies depending on which level the situation is viewed from. Generally, the last user, which is the building, expects to reduce the energy consumption costs or extend the duration of the benefit from an energy source. These goals should be achieved with a minimum payback period or with maximum cost savings. The answer for an energy distribution company or a government is to reduce the peak load. However, it is well known that the most of ES systems always provide lower energy consumption costs and contribute positively to sustainability indexes.

1.6 History of Energy Storage

Energy has been an essential need for people since ancient times. While energy had been thought of coming from food sources for people's activities such as walking, running, hunting, and so on, energy had been started to be derived from harboring, heating, cooling, and so on with civilization and communal life in ancient times. With the start of mechanization, people had begun to benefit from different energy sources such as rivers, winds, and animal power to meet their different energy needs. Fig. 1.23 shows some ancient machines powered by wind, water, and animal. In Fig. 1.23a, the first crank system, design by El-Cezeri, had been used for pumping water from lower level to upper level. In Fig. 1.23d and e, the animal-driven and wind-driven systems had been used for producing flour. These systems had been used just when the energy source had been active. The energy needs have increased with the increasing population, and these increased needs have led people to search for different energy sources and different energy conversion systems.

With the industrial revolution and the invention of the steam engine, energy needs have been moved to a very different dimension. People have stored the fuels used in the steam engines such as coal, charcoal, wood, etc., for later usage when the energy has needed. The invention of electricity and the increase in people's energy needs have forced people to investigate the new energy sources and energy conversions system. We have started to use the power plants run with the Rankine cycle and Brayton cycle. We have also started to use nuclear and renewable energy sources. The reason for using so many different energy sources in those days was to meet the increasing need for energy. Increasing energy demand is meant for more power plants. And people have started to use many energy sources from nuclear to solar energy, from wind to coal, and from fuel to natural gas. The perpetual increase in energy demand has led people to make more efficient use of energy resources, and the concept of energy efficiency has become an important concept for people. The concept of energy efficiency has paved the way for multigeneration systems and energy recovery systems. Within this period, ES systems have come to the forefront as an important tool both in making more efficient use of energy resources and in reducing energy losses. Additionally, increasing energy production and demand have caused an imbalance between them. ES has also played a key role in solving this mismatch problem.

FIG. 1.23

Some ancient examples about energy storage over centuries.
(a) Adapted from Ref. [21]; (b) Adapted from Ref. [24]. (c) Adapted from Ref. [22]. (d) Adapted from Ref. [23].
(e) Adapted from Ref. [25].

The oldest known ES application is harvesting the ice and snow during the winter for later use. In Fig. 1.23c, an icehouse from Italy is shown. Icehouses are used as refrigerator to keep the foods for later usage. In ancient times, to store energy sources had been assumed as ES, such as wood, food, coal, and so on. Solar heating systems and dams are the oldest ES methods. The practical ES application has been started in the 19th century. The world's first hydroelectric power plant, shown in Fig. 1.23b, has been built in USA in 1882. The first industrial solar heating system was developed in the 19th century. Also, the first battery, which was Volta's cell, was developed in the early 1800s. Flywheels and compressed springs were the first mechanical ES systems. Heat storage and its application have been taken place in the practical application in a form of hot water tanks, solar ponds, and rock beds. Due to the increasing energy demands, and requirement in the use of energy in an optimum way, ES techniques are a significant technology. There are many methods and applications for ES. They have been discussed in the next section.

1.7 Closing Remarks

In this chapter, first, energy consumption and its environmental impacts have been discussed to reveal the need for ES systems. Then, ES methods have been introduced to readers. Next, the advantages of ES have been explained with practical examples. Lastly, the history of ES has been presented briefly to the readers. It is clear that ES methods are a significant solution for managing the energy systems in an optimum and sustainable way. Integrating the ES systems into energy conversion systems can reduce energy consumption and costs substantially. Also, they provide flexible working conditions for energy systems. They are a unique option to manage grid demand and supply profiles.

References

[1] BP, Statistical Review of World Energy, 2020. https://www.bp.com/en/global/corporate/energy-economics/statistical-review-of-world-energy.html. (Accessed 26 April 2020).

[2] GridWatch, GB Electricity National Grid Demand and Output, 2020. https://gridwatch.co.uk/. (Accessed 4 January 2020).

[3] V. Smil, Energy Transitions: Global and National Perspectives, second ed., Praeger, 2016.

[4] World Energy Outlook 2019, 2019. https://www.iea.org/data-and-statistics. (Accessed 4 January 2020).

[5] I. Dincer, Energy and environmental impacts: present and future perspectives, Energy Sources 20 (1998) 427–453.

[6] I. Dincer, C. Acar, A review on clean energy solutions for better sustainability, Int. J. Energy Res. 39 (2015) 585–606, https://doi.org/10.1002/er.3329.

[7] I. Dincer, Smart energy solution, Int. J. Energy Res. 40 (2016) 1741–1742, https://doi.org/10.1002/er.3621.

[8] I. Dinçer, M.A. Rosen, Thermal Energy Storage: Systems and Applications, second ed., 2010, https://doi.org/10.1002/9780470970751.

[9] C. Cheng, A. Blakers, M. Stocks, B. Lu, Pumped hydro energy storage and 100 % renewable electricity for East Asia, Global Energy Interconnect. 2 (2019) 386–392, https://doi.org/10.1016/j.gloei.2019.11.013.

[10] B.S. Zhu, Z. Ma, Development and prospect of the pumped hydro energy stations in China, J. Phys. 1369 (2019) 012018, https://doi.org/10.1088/1742-6596/1369/1/012018.

[11] Hydrostor, n.d. https://www.hydrostor.ca/applications/. (Accessed 1 May 2020).

[12] D. Erdemir, I. Dincer, Assessment of renewable energy-driven and flywheel integrated fast-charging station for electric buses: a case study, J. Energy Storage 1 (2020) 1.

[13] M.S. Guney, Y. Tepe, Classification and assessment of energy storage systems, Renew. Sustain. Energy Rev. 75 (2017) 1187—1197, https://doi.org/10.1016/J.RSER.2016.11.102.

[14] Hydrogen Storage, Office of Energy Efficiency and Renewable Energy, 2020. https://www.energy.gov/eere/fuelcells/hydrogen-storage. (Accessed 5 January 2020).

[15] A. Züttel, A. Borgschulte, L. Schlapbach, Hydrogen as a Future Energy Carrier, Wiley-VCH Verlag GmbH & Co. KGaA, 2008, https://doi.org/10.1002/9783527622894.

[16] T. Wetherell, Ammonia-based Energy Storage, n.d. https://cecs.anu.edu.au/research/research-projects/ammonia-based-energy-storage. (Accessed 13 July 2020).

[17] J. Pospíšil, P. Charvát, O. Arsenyeva, L. Klimeš, M. Špiláček, J.J. Klemeš, Energy demand of liquefaction and regasification of natural gas and the potential of LNG for operative thermal energy storage, Renew. Sustain. Energy Rev. 99 (2019) 1—15, https://doi.org/10.1016/j.rser.2018.09.027.

[18] Electrochemical Energy Storage: Simple Definition, n.d. https://www.energie-rs2e.com/en/articleblog/electrochemical-energy-storage-simple-definition. (Accessed 5 February 2020).

[19] R.F. Service, Advances in flow batteries promise cheap backup power, Science 362 (2018) 508—509, https://doi.org/10.1126/science.362.6414.508.

[20] M.G. Molina, Dynamic modelling and control design of advanced energy storage for power system applications, Dyn. Model. (2010) 49—92, https://doi.org/10.5772/7092.

[21] S. Atabey, El-Cezeri, 2014. https://komhedos.com/el-cezeri/. (Accessed 13 July 2020).

[22] I. Sailko, Ice House (Building), n.d. https://commons.wikimedia.org/w/index.php?curid=2823200. (Accessed 13 July 2020).

[23] V. Labate, Roman Mills, 2016. https://www.ancient.eu/article/907/roman-mills/. (Accessed 13 July 2020).

[24] The World's First Hydroelectric Power Plant, n.d. http://www.americaslibrary.gov/jb/gilded/jb_gilded_hydro_1.html#:~:text=On September 30%2C1882%2C the,by Appleton paper manufacturer H.J. (Accessed 13 July 2020).

[25] Nashtifan Windmills, n.d. https://www.atlasobscura.com/places/nashtifan-windmills. (Accessed 13 July 2020).

CHAPTER

Heat Storage Methods

2

2.1 Introduction

Energy storage is no longer an option, but is a compulsory requirement to make energy systems more feasible, more cost-effective and more environmentally friendly. This is not enough! It also helps offset the mismatch between demand and supply and overcome the fluctuating nature of renewable energy sources. Although some introductory details are presented in Chapter 1 to discuss the details about energy storage methods and their utilization in various sectors, this particular chapter is aimed to focus on heat storage methods. Heat generally refers to thermal energy. People use both interchangeably, and both are correct. However, some people use heat storage as a method under thermal energy storage, which is not thermodynamically correct. That is why, we need to have correct terminology to use throughout. This is also very important to rightly guide the researchers, scientists, engineers, and practitioners.

Although there are multiple energy storage methods practically applicable to various systems and applications, thermal energy (heat) storage methods, in particular sensible and latent type heat storages, are the most commonly deployed energy solutions for various heating, cooling, and air-conditioning (AC) applications. More than half of consumed and generated energy around the world occurs in the form of heat. Almost all losses and irreversibilities occur as heat in all systems. For these reasons, heat storage techniques have significant potential to manage the energy systems in an optimum way and help reduce and/or recover the losses for useful outputs.

Heat storage, also called thermal energy storage, is to store energy temporarily by converting an energy type to heat for later usage. The stored energy can be used as heat energy, or it can be converted into another type of energy. Heat storage deals with the storage of energy by cooling, heating, and phase changing (solidifying, melting and vaporizing) of a material. The stored energy becomes available when the process is reversed. Heat storage techniques are one of the most commonly used and mature and oldest energy storage methods. Heat storage systems are also one of the most critical energy storage techniques since heat is a last-user type of energy. Thus, the stored energy can be used directly without converting a type of energy.

Energy demands in the industrial, commercial, and residential sectors alter significantly on daily, weekly, seasonal, and yearly bases as pointed out in Figure 1.2. The daily energy demand peaks when the energy demands in all sectors coincide. On a weekly basis, since most of the utility and industrial demands stop at weekends, energy demands decrease substantially. Energy demands on a yearly basis depend on seasonal energy demands such as heating and cooling. These fluctuating energy demands can be balanced with the help of heat storage techniques. Numerous heat storage systems have been developed in industrialized countries in the past half-century. Heat storage systems allow equipment used in thermal systems to be used more efficiently, with lower capacity and/or less operating costs. Due to these benefits, heat storage systems have significant potential in terms of economy and environment.

Heat storage techniques have an increasing interest due to thermal system applications such as space heating—cooling, water heating—cooling, and AC. Heat storage techniques have a great potential to use system equipment more efficiently, to reduce the capacity of system equipment, and to shift the time-of-use periods of energy consumption. It is one of the most important methods that can solve the problem of time matching between when energy is needed and when it is produced. In addition, it also supplies to drive an energy conversion system in a less costly manner by taking advantage of the tariff difference in the unit price of electricity.

The objective of this chapter, first, is to define the terms of thermal energy, heat, and temperature. Then, heat storage methods, their classifications, and applications are given to readers. The advantages and objectives of heat storage systems are discussed over the practical applications. Finally, phase change materials (PCMs) used in the heat storage systems are introduced and discussed for various systems and applications.

2.2 Temperature, Heat and Heat Capacity

Temperature, thermal energy (so-called heat), and heat capacity are closely relevant and significant terms for energy conversion and transportation systems. They are often confused with each other and used instead of someone. Therefore, first, the terms of temperature, thermal energy, and heat are going to be defined for readers. Although thermal energy and heat are thought different terms, actually, thermal energy and heat refer to the same thing. However, in many books and papers, it is assumed that heat is related to the transfer of thermal energy from one substance to another or the environment depending on the temperature difference. In other words, heat is the transferred form of thermal energy.

Thermal energy and temperature are concerned with the kinetic energy of molecules. While thermal energy is the sum of kinetic energies of all molecules of a substance, the temperature is a measurement of the average kinetic energy of molecules of a substance. Temperature is a physical property of matter that quantitatively expresses hot and cold. Fig. 2.1 schematically describes the concepts of thermal

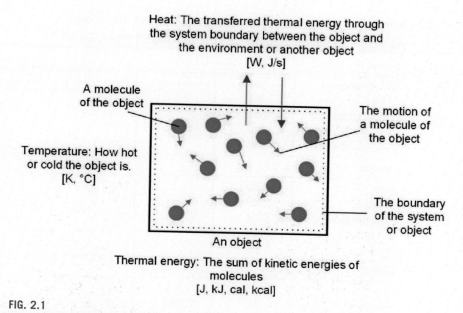

Heat: The transferred thermal energy through the system boundary between the object and the environment or another object [W, J/s]

A molecule of the object

The motion of a molecule of the object

Temperature: How hot or cold the object is. [K, °C]

The boundary of the system or object

An object

Thermal energy: The sum of kinetic energies of molecules [J, kJ, cal, kcal]

FIG. 2.1

A schematic view of thermal energy, heat, and temperature for an object.

energy, heat, and temperature for an object. Thermal energy is the energy content of an object or substance in a particular condition and measured in joules (J), calories (cal), and British thermal units (Btu). It is also defined as the total microscopic kinetic and potential energy of a substance. Temperature is a measurement of the grade of thermal energy for a substance. While higher thermal energy content means higher temperature, low thermal energy content means a lower temperature. Temperature is measured in Kelvin (K), Celsius (°C), Fahrenheit (°F), or Rankine (R). It is a measurable physical property of an object and also known as a thermodynamic state variable. Heat is the transferred thermal energy from an object to the environment or another object in contact, or vice versa. The unit for the rate of heat transfer is the watt (W), defined as 1 J per second.

The heat required (Q) to heat a volume (V) of a substance from a temperature (T_1) to a temperature (T_2) is calculated by

$$Q = mc(T_2 - T_1) = \rho VC(T_2 - T_1) \qquad (2.1)$$

where c is the specific heat of the substance, m is the mass of the substance, and ρ is the density of the substance. The specific heat is the amount of heat per unit mass required to raise the temperature by 1°C. It is also the relationship between thermal energy and temperature change. For instance, two different materials have different temperatures depending on specific heat values even if they have equal thermal energy content. In other words, when the same amount of heat is transferred to two different materials, the temperature changes in these materials will be different.

Table 2.1 The Specific Heats of Various Common Materials.

Material	c_p [J/kg K]	c_v [J/kg K]	Temperature
Solids			
Aluminum (pure)	896		20°C
Iron (pure)	452		20°C
Copper (pure)	383		20°C
Silver (99.9% pure)	234		20°C
Nickel (99.9% pure)	446		20°C
Liquids			
Water	4180		25°C
R134a	1340		20°C
Ethylene glycol	1109		20°C
Benzene	1720		20°C
Mercury	139		25°C
Gases			
Air	1005	718	300 K
Hydrogen	14,207	10,183	300 K
Steam	1872	1410	300 K
Nitrogen	1039	743	300 K
Carbon dioxide	846	657	300 K

Data from Ref. [1].

While higher specific heat values mean lower temperature change, lower specific heat values mean higher temperature change. There are two kinds of specific heat: specific heat at constant volume (c_v) and specific heat at constant pressure (c_p). The specific heats of some common types of materials are given in Table 2.1.

2.3 Sensible Heat and Latent Heat

Heat change in a material occurs in two different types: sensible and latent. If the energy is released from a material, the temperature of that material will decrease. Otherwise, when a material absorbs the energy, the temperature of the material increases. That heat change in the material is called the "sensible heat." Latent heat is concerned with the phase change of a material. For instance, while energy is required to convert ice to water, to change water to steam, and to melt the paraffin, the energy will be released in the phase change from water to ice, steam to water, and melted paraffin to solid paraffin. When the material is pure, the temperature of the material stays constant during the phase change. The energy required to cause these changes is called the "heat of fusion" at the melting point and the "heat of

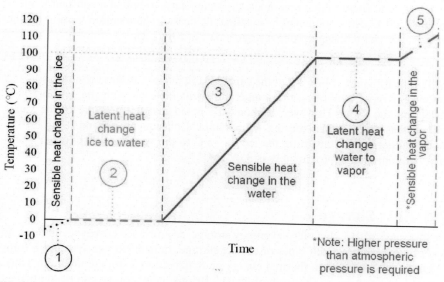

FIG. 2.2

A temperature variation diagram for the phase change of water from solid to vapor.

vaporization" at the boiling point. While the sensible heat is relating to the specific heat of a material, the latent heat is concerning with the heat of fusion and heat of vaporization.

For example, let us consider water (H_2O), and suppose that we wish to obtain vapor from a 1 kg of ice block at a temperature of $-5\,°C$ and the pressure of 1 atm (101.325 kPa). Fig. 2.2 shows the temperature change of the water during phase change. First, energy is required to increase the temperature of the ice block, and the temperature of the ice block will increase up to 0 C with given energy to the ice. Assuming the specific heat of the ice is 2100 J/kg K, the required heat can be calculated as

$$Q_1 = mC(T_1 - T_0) = 1 * 2100 * (0 - (-5)) = 10,500\ J \tag{2.2}$$

Then, the ice starts to melt at a constant temperature of 0°C, as seen in the second part of the diagram. During the ice melting, the energy change is the latent heat. Latent heat during the phase change of ice/water is calculated as follows:

$$Q_2 = m\,h_{fusion} = 1 * 333500 = 333,500\ J \tag{2.3}$$

Next, after the ice/water converting is completed, the temperature of the water starts to increase with the entering energy content into the water. Again, the sensible heat conditions are valid, as seen in the third part of Fig. 2.2. The required energy is calculated:

$$Q_3 = mC(T_3 - T_2) = 1 * 4180 * (100 - 0) = 418,000\ J \tag{2.4}$$

Last, if the energy input continues, the water starts to boil at a constant temperature of 100°C. Now, the latent heat called the heat of vaporization is valid for the water vapor phase change process (fourth part of Fig. 2.2).

$$Q_4 = m\, h_{\text{vaporization}} = 1 * 2260,000 = 2,260,000 \text{ J} \tag{2.5}$$

For the fifth part of Fig. 2.2, if the vapor is kept in the constant volume vessel and the energy input continues, the temperature will increase with the sensible heat change of the vapor. However, it should be noted that the pressure of the vapor will increase with the increasing temperature. Consequently, from T_0 to T_4, the required energy for converting the ice at −5 C to the vapor at 100 C under 1 atm pressure condition approximately is 3.022 MJ.

2.4 Heat Storage

Heat storage, also called thermal energy storage, as advanced energy technology, is one of the most practiced types of energy storage. In heat storage systems, a form of energy is converted to thermal energy and charged to a storage medium. Thus, energy can be stored in the form of "heat." Briefly, heat storage can be defined as heat change in a medium. The illustration of heat storage for an object is seen in Fig. 2.3. In a typical heat storage system, an energy source is converted to heat and charged to a storage medium. The heat charged to the storage medium causes the temperature change or phase change. Temperature change or phase change in the storage medium is the stored heat. The stored energy is used by performing reverse temperature change and phase change during the discharging period. A heat loss occurs during the charging, storing, and discharging periods. Minimizing heat losses is recognized as a critical issue for the heat storage system to increase system performance.

The heat storage capacity for a storage medium directly depends on mass, specific heat, and heat of fusion according to sensible and latent heat storage applications. When sensible heat storage is considered, the heat storage capacity

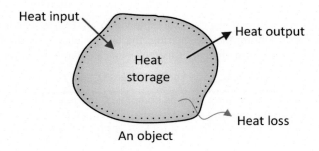

FIG. 2.3

An illustration of heat change in an object.

is determined with the mass and specific heat of the storage medium. The increasing mass of storage medium increases the heat storage capacity. However, it is not possible to continuously increase the mass of the storage medium. The mass is generally limited to physical conditions. Therefore, the use of the storage mediums that have a higher specific heat provides a lower mass for a given storage capacity. In Table 2.1, the specific heats of various common materials are given. In latent heat storage systems, the heat of fusion is a critical parameter besides the mass of the storage medium. Higher heat of fusion provides higher storage capacity for a given storage volume or mass. Also, the latent heat change is always higher than the sensible heat change for a storage medium. An example of sensible and latent heat change for water is given in Section 2.3. Also, it is discussed in the following sections in detail.

Heat storage systems can be classified into two different categories: sensible and latent. There are also hybrid heat storage systems that combine sensible and latent heat storage methods. Heat storage techniques are going to discuss in the following sections. In a number of sources, heat storage methods are classified into three different categories: sensible, latent, and thermochemical heat storage methods. In the sensible and latent heat storage techniques, the form of the stored energy is the heat. However, in the thermochemical heat storage method, energy is stored in a chemical compound. Forms of energy inlet and output are the heat. During the charging period, an endothermic reaction is performed, and a chemical compound is obtained. This chemical compound is stored in a container. During the discharging period, the stored chemical compound is used in an endothermic chemical reaction, and energy is released. In this book, the thermochemical heat storage is included in chemical energy storage methods.

Heat storage systems are applicable in several residential and industrial purposes. Heat storage can be used for storing any forms of energy. For instance, solar energy is stored in solar domestic hot water systems or solar ponds in the form of sensible heat. The electricity load depending on AC systems can be stored in the cold thermal energy storage systems. Additionally, heat storage can be used for the organic Rankine cycle (ORC) for electricity storage. Heat storage techniques are a significant option to manage and improve the energy conversion systems such as space heating, hot water, and AC. Heat storage systems are provided to operate these systems in a more efficient and sustainable way. Therefore, as advanced technology, heat storage techniques play a significant role in reducing energy costs and device capacities. However, the objectives of heat storage are not limited to energy costs and device capacities. They have many advantages depending in their objectives.

2.4.1 Purposes of Heat Storage Techniques

Heat storage is basically defined as the storage of any forms of thermal energy by converting to heat in a storage medium for later use. The objective of later use may aim to achieve the following:

- to continue to utilize the energy source when it is not active;
- to reduce energy consumption costs;
- to reduce the peak energy loads;
- to shift the peak energy loads to off-peak periods;
- to reduce the size and capacity of the system equipment;
- to balance energy supply and demand;
- to minimize and regain the energy losses; and
- to use during emergencies.

A heat storage system may provide one, a few or all of these benefits depending on the utilization strategy of the heat storage system. These aims are also several advantages of heat storage systems. These aims are set in accordance with the heat storage utilization strategy. To obtain the desired aim, the proper heat storage utilization strategy should be applied to the system. The heat storage utilization strategy has been discussed in the following section. Now, the aims of the heat storage listed earlier are explained with practical examples.

Continuing to utilize the energy source when it is not active: This aim is also called as solving the mismatch problem between the energy supply and demand. Heat storage—integrated solar heating systems are a basic example of continuing to utilize the energy source when it is not active. In solar energy systems, solar energy is stored in a storage medium (water, rock bed, solar pond, PCM) to use when solar energy is not active during the night hours or when solar energy is lower during the winter. Then, energy demand is met by the stored energy. Solar domestic hot water systems are one of the typical and practical applications of short-term storing of solar energy. In these systems, solar energy is converted to heat and then transferred to a heat transfer fluid (air, water, or glycol ethylene solutions). The heat transfer fluid goes to a storage tank and increases the temperature of the water in the tank. This process continues as a cycle during the day. The obtained hot water in the storage tank is used at night. The solar-driven heat storage in a rock bed can be given as an example of high-capacity and long-term storage of solar energy. Solar-driven borehole heat storage systems are also one of the practical applications of high-capacity and long-term heat storage. The working principle of these systems is similar to a solar domestic hot water system. Solar energy is charged to a rock bed or borehole by converting to heat through a heat transfer fluid during the summer. The stored sensible heat is used to meet the energy demands during the winter.

Reducing energy consumption costs: Heat storage systems reduce energy consumption costs in two ways. First, since they extend the use period of energy sources especially for renewable energy sources, energy consumption costs can be reduced by the heat storage systems. Second, it can reduce energy costs by taking advantage of the change of energy tariff during the day. It is a common application to divide the day by three, which are off-peak, midpeak, and peak periods, in terms of electricity time of use periods to promote lowering the peak energy demands. Fig. 2.4 illustrates the shifting cooling loads from expensive periods to cheap periods. The red zone in Fig. 2.4 is the most expensive period called the peak period. Cooling loads

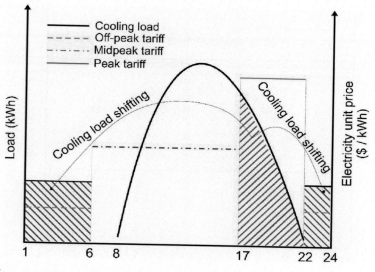

FIG. 2.4

Demonstration of the strategy of reducing energy consumption costs by shifting the loads in electricity peak hours to off-peak hours.

in the peak period can be shifted to the off-peak period (blue zone), which is the lowest unit energy price by heat storage methods. Thus, energy demands can be met with a cheaper solution thanks to heat storage systems. Ice storage systems, also called ice thermal energy storage, are an innovative method to reduce the operating cost of AC systems. In ice storage systems, ice is produced during the electricity off-peak hours (in the cheapest electricity unit price). The cooling needs of the building, when the electricity unit price is higher, are met with the stored ice. Thus, the operating costs of ACs can be reduced substantially.

Reducing the peak energy loads: Energy conversion devices in buildings are designed to meet peak loads. However, in a typical building, the peak energy demand is seen in a very limited time when the daily or seasonal energy demand period is considered. High-capacity energy conversion devices are used for meeting the peak energy demands. High-capacity devices are also meant a high initial cost. To essentially meet the high- but short-period, peak load is required a substantial amount of investment. In addition, this high capacity brings high power requirements. High-capacity energy conversion devices also increase the capacities and costs of auxiliary devices used in the building, such as generator, cable, and transformer. Heat storage systems are a significant solution to reduce the peak load, capacity of devices, and power requirement. The load leveling strategy, shown in Fig. 2.5, is used for reducing peak loads and parameters associated with it. Thus, the peak load is minimized. Besides, the total energy demands are met with lower costs since a portion of the energy demand is shifted to off-peak hours. To reduce the peak energy loads and to distribute the energy loads equally during the day,

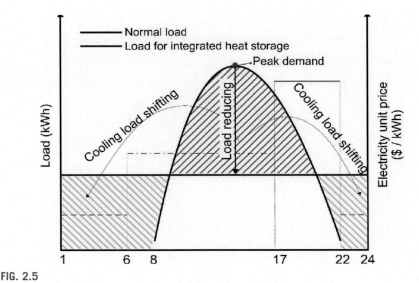

FIG. 2.5

The load leveling strategy.

the load leveling strategy is used in heat storage systems. The load leveling strategy provides the lowest capacities of energy conversion and auxiliary devices.

Shifting the peak energy loads to off-peak periods: Heating and cooling systems are responsible for more than half of the energy consumption of residential buildings. Heat storage systems play a critical role to balance all energy demands of buildings by shifting heating and cooling loads to lower demand hours. Fig. 2.6 shows the energy loads for a typical office building. It is clear from Fig. 2.6 that the cooling load is responsible for approximately 40% of all energy loads. The cooling loads can be shifted to lower energy demand hours or electricity off-peak tariff hours. Thus, both the cooling cost can be reduced, and the energy

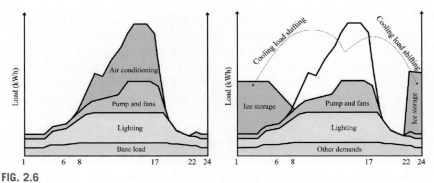

FIG. 2.6

Hourly energy loads for a typical office building with and without heat storage.

demand can be balanced. As discussed before, while shifting the peak loads to off-peak hours balances the energy demands, the energy consumption costs reduce considerably. This aim is an important option to reduce the need for building power demand.

Reducing the size and capacity of the system equipment: The power requirement of the buildings is a significant issue since it is related to the grid load. All buildings are designed for their peak load conditions. However, every day, a new need item is entering our lives. For example, while AC systems were not common a few decades before, now, they are an important need for people's thermal comfort. Plug-in electric vehicles are also a good example of new energy needs. Most of the buildings do not have a proper system for charging the cars and also the infrastructure for charging purposes because the electric requirements for charging vehicles were not included in the building peak energy demands. Due to ever-increasing use of electric vehicles, this demand increases day by day. If the infrastructure of the building is not convenient for charging vehicles or there is no potential grid load for charging, proper power demand can be adjusted by using heat storage systems. The peak load can be reduced by heat storage systems, and the obtained load gap can be used for meeting the charging or other new demands.

Reducing and regaining heat losses: Heat storage systems may also be used for heat recovery. Most devices used in the buildings cause heat losses when they operate. Heat recovery can be performed simultaneously while the system is working, or it can be stored for later usage in heat storage systems. Thus, it is possible to benefit from the recovered heat for a longer time. This reduces the losses in the system and increases the useful output. For example, the waste heat in the ORC can be stored for later use for heating purposes.

Using during emergencies: In buildings where cooling is important and critical, such as data-processing centers, supercomputer centers, stem cell and embryo centers, and museums, there are one or more substitute devices for cooling systems and their generators. This brings along with it a substantial initial investment and service costs. Ice storage or other cold heat storage techniques can be used for emergencies. Thus, both cooling costs and the costs of devices used for emergencies can be reduced.

It is important to note that the heat storage methods used for the different purposes listed earlier during normal periods can be used to meet the extreme burden that occurs in emergency situations. During the COVID-19 pandemic, the capacities of hospitals have been exceeded substantially around the world. However, the capacities of AC systems in hospitals are not enough to meet this unexpected demand. While cold heat storage systems can be used for meeting the unexpected high cooling demands during emergencies, they can reduce the cooling costs and shift the peak loads to off-peak hours during the normal days.

2.4.2 Heat Storage Utilization Strategy

The heat storage utilization strategy is basically defined as how much and when energy is stored, and how and when to be used the stored energy. The heat storage utilization strategy is also called as the storage scenario. Heat storage utilization strategy is directly related to its objectives defined in the previous section. The specifying of the heat storage utilization strategy is one of the most important issues for an energy system—integrated heat storage energy system since it depends on many criteria such as tariff structure and the physical condition of the plant. It has a significant impact on the system performance indicators such as the savings, payback period, efficiency, amount of the storing energy, and so on. Therefore, it should be determined and applied carefully. The selection of the heat storage utilization strategy is done according to the following criteria:

- The cost involved in energy use
- The environmental impact in terms of primarily carbon dioxide emissions
- The load profile of energy demand
- The type of energy source
- The utilization time of energy source
- The capacity requirement and limitations about the facility
- The tariff structure of energy source
- The objective of heat storage medium (such as reducing energy costs, shaving the peak loads, and reducing sizes of the system devices)

First, the amount of energy to be stored should be determined since heat storage systems are required a storage volume for a storage material and its container. Heat storage capacity generally depends on the physical conditions of the facility, and it is limited by the volume available for the storage container or medium. To increase storage capacity, underground heat storage tanks are used in practical applications. If the storage volume has a natural structure such as rock bed, pond, and so on, the storage capacity is limited by the volume and capacity of the medium. If the storage volume has a manufactured structure such as tank and vessel, it is limited by the structure of the facility for the placement of the storage container. If the heat storage system is included in the construction plan of the building, it is possible to arrange the optimum storage capacity. If the heat storage system is integrated into the building energy system for an existing building, it may be difficult to adjust the volume required for the targeted storage capacity. The storage container can be placed underground or in a zone to be separated from the building, such as the parking lot or basement.

The type of energy source is also an important criterion for heat storage system. The hours when the energy source is active is usually used as the charging period. The stored energy can be kept in the storage medium during the storing period. When the energy source is not active, the stored energy is used for meeting energy needs. This case is generally seen in renewable energy sources. For instance, in the solar-driven systems, the charging period is performed during the day time or summer. In the wind-driven systems, the determination of the charging period has been complicated due to its unstable structure.

The energy demand profile also called energy load change over time is one of the significant criteria for selecting a heat storage utilization strategy. It also determines

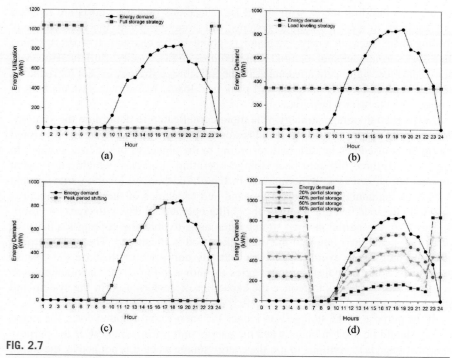

FIG. 2.7

The heat storage utilization strategies: (a) full storage, (b) load leveling, (c) peak load shifting, and (d) partial storage.

Data from Ref. [2].

the objective of heat storage given in Section 2.4.1. Practically, in terms of the amount of storage, heat storage systems are basically divided into two: full and partial storage. In the full storage strategy, all energy demand is stored by the heat storage systems. The discharging period covers all energy demand periods. The charging period is determined according to the availability or unit price of the energy source. Fig. 2.7 demonstrates the cooling load distribution of the heat storage utilization strategies for AC systems used in a hypermarket. As can be seen from Fig. 2.7, the hypermarket and its AC system are active between 08:00 and 22:00 during the summer season. Hypermarkets have generally high cooling loads due to high people circulation and fresh air requirements. Therefore, the operating AC requires a significant cost for hypermarkets. The ice storage system also called ice thermal energy storage and heat storage in ice is an essential option to reduce cooling costs and manage the capacities of devices used in AC systems. When all cooling demands meet with the AC-integrated ice storage system (full storage strategy, Fig. 2.7a), chillers are operated during the off-peak electricity hours, and ice is produced. During the day, mostly midpeak tariff hours, the cooling need of the building meets with stored ice. Thus, AC operating costs are reduced substantially. In most ice storage systems, full storage capacity increases the peak cooling load since all energy demand is going to be charged in a shorter time than the normal demand period. This requires high-capacity chillers. However, one of the most significant advantages of the full storage strategy except for reducing cooling costs is to shift midpeak and peak grid loads. Ice storage systems help to manage the electricity grid loads.

In the partial storage strategy, it is more complicated to determine the strategy. While a portion of the energy demand is stored by the ice storage system, how stored energy to be used can be different according to the objective of the heat storage. In Fig. 2.7b–d, three different heat storage operating strategies, which have partial storage, are shown for a hypermarket. In Fig. 2.7b, the load leveling strategy that is used for reducing peak cooling demand and capacities of the AC equipment is shown. In the load leveling strategy, the chiller operates all day with constant power. In Fig. 2.7c, the partial storage strategy that aims to shift the cooling loads in the electricity peak tariff hours to the off-peak period is illustrated. The main purpose of this strategy is to avoid expensive electricity unit price or to shave the electric peak loads. Some partial storage strategies are shown in Fig. 2.7d. The main purpose of this partial storage is to reduce the capacities of devices used in the system and energy consumption costs.

If the purpose of the storage is to reduce the energy consumption costs, the stored energy should be used in hours when the energy unit price is highest. If the purpose of heat storage is to continue to use the energy source when it is not active, the stored energy should be used in hours when the energy source is not available. Or the stored energy may be used for meeting the energy demand during the grid energy peak periods in order to reduce the peak grid loads and to balance energy supply and demand profiles. Consequently, the heat storage utilization strategy should be carefully determined, since it directly affects the initial cost, savings, and payback period. The effect of utilization strategies on the economic indicators is discussed in the case studies.

2.4.3 Working Principles of Heat Storage Systems

The basic principles of heat storage techniques are the same as other forms of energy storage methods. A typical heat storage system consists of charging, storing, and discharging periods. These periods run as a cycle according to the strategy of storage. Fig. 2.8 demonstrates a basic storage cycle for a heat storage and cold heat storage. The duration of storage steps varies with the heat storage utilization strategy. In some storage systems, some of the steps may occur synchronously, or a step may run more than once in each storage cycle.

In the charging period, an energy source is converted to heat, and it is used for changing the temperature of storage material or changing the phase of storage material. The charging period is performed when the energy source is either active, cheap, or nonintensive use time. For example, in solar energy systems, the charging period is performed during the daytime or summer when the solar energy intensity is

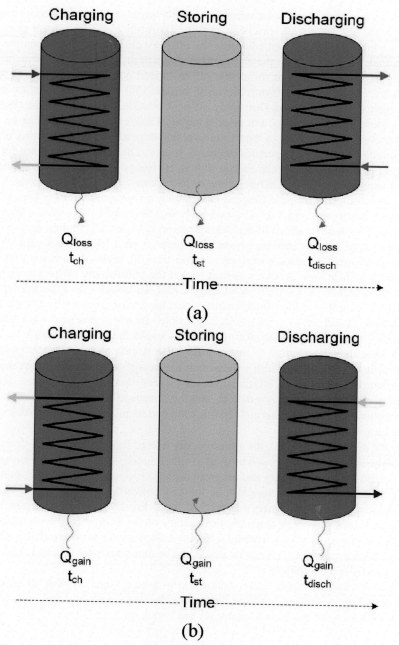

FIG. 2.8

Three processes for (a) a heat storage system and (b) a cold heat storage system: charging (left), storing (middle), and discharging (right).

higher. To continue to use solar energy during the night or winter season, it can be stored in the hot water tank, rock beds, or melted paraffin. In the ice storage systems, the charging period is performed during the off-peak electricity hours when the use of electricity is lower. Thus, the electricity peak loads are shifted to off-peak hours, and cooling cost is reduced. In almost all developed countries, time-of-use tariff structures, called triple tariff or multi tariff, are used for promoting to reduce peak loads and to balance the load. During the off-peak periods, the energy can be charged for later use when the energy unit price is higher. Thus, energy demands can be met at lower costs. The duration of the charging period depends on the availability of energy source or time-of-use periods of energy source. The most critical point for the charging period is to charge the energy with an efficiency as high as possible since it can be supplied the lower charging time or lower capacity charging devices.

In the storing period, the charged energy into a storage medium is kept in the container (such as tanks, vessels, ponds, beds, etc.) until the stored energy is required. The main problem in the storing period is losses. Depending on the temperature of the storage medium, losses can occur as heat loss or heat gain. When the temperature of the storage medium is higher than the environmental temperature, it happens as heat loss. It occurs as heat gain if the temperature of the storage medium is lower than the environmental temperature. To reduce losses, generally, heat insulation is applied to the storage medium container. Also, the duration of the storing period should be as minimized as possible. In many practical heat storage methods, the efficiency of the storing period is higher than the other periods.

In sensible heat storage systems during the storing period, the temperature gradient in the storage medium is important to manage the losses and heat storage performance. Temperature stratification also called thermal stratification defines the degree of the temperature gradient, and the improved thermal stratification provides better heat storage. Thermal stratification and its impacts have been discussed in the case studies.

In the discharging period, the stored energy is used for meeting energy demands. While the discharged energy can be used to either directly meet the energy demands, or it can be used for auxiliary and preprocess purposes, such as preheating, precooling, and so on. Besides, it may also be used by converting to any form of energy. Consequently, the stored energy is used to meet directly the energy demands, or to reduce loads of energy demands or capacity of the energy conversion devices. The discharging period is performed when the energy source is not available, expensive, or peak hours. Thus, energy demands can be met with more efficient, low-cost, or reduced-sized equipment.

Heat storage processes may require extra devices such as heat exchangers, pumps, and so on. This requirement depends on the heat storage method, storage material, packaging of the storage material, and container.

2.4.4 Benefits of Heat Storage

When the energy needs of people are taken into consideration from past to present, the heat has been the first and important one. Today, electricity is the first energy need in the world. Usually, electricity comes to mind, when the energy is told. However, electricity mostly converts to heat as part of the energy conversion processes or losses. Or, electricity often is generated via a thermal cycle such as Rankine cycle, Brayton cycle, and so on. Therefore, thermodynamics, heat, and heat transfer are a critical topic for energy conversion and generation processes. Since heat is significant for the energy conversion and generation processes, heat storage techniques can play a critical role to increase the flexibility of the process, to recover losses, and to increase the useful output. Heat storage methods have many advantages like other energy storage methods. The benefits of heat storage methods are listed in the following:

- To reduce the energy consumption costs
- To improve the efficiency of the system's devices
- To extent the period of use of energy source
- To decrease the capacity and size of the system's equipment
- To reduce the initial investment and service costs
- To balance the energy supply and demand
- To increase the generation capacity
- To improve the flexibility of working conditions and hours
- To reduce the fossil-based energy consumption
- To minimize the carbon emission and environmental impact

All listed benefits are very critical to meet energy demands in a sustainable way. They are also significant to managing the energy supply and demand profiles for individual users, energy-generating and distribution companies, and governments. A heat storage system may provide one, a few or all these benefits depending on the utilization strategy of the heat storage system. Someone may ask: What is the most significant benefit to be provided by a heat storage system? The answer will be different for the level (individuals, companies, or governments), which monitored the benefit of heat storage. Generally, the last user, which is the building, expects to reduce the energy consumption costs or extend the duration of the benefit from an energy source. The main aim of heat storage is to provide a minimum payback period or maximum cost savings for the last users. The answer for an energy distribution company or a government will become to reduce the peak load. However, it is well known that heat storage systems always provide lower energy consumption costs and contributed positively to sustainability indexes. Therefore, incentive practices are being carried out by governments and electricity distribution companies to spread the use of heat storage systems.

2.5 Heat Storage Methods

Heat storage has two methods: sensible and latent heat storage methods. The use of heat storage systems in building applications is for mostly heating and cooling purposes. Different energy sources are used for heating and cooling of the buildings. Therefore, to store heating and cooling capacities, the different energy sources are

Table 2.2 Available Media for Sensible and Latent Heat Storage Systems.

Sensible Short Term	Latent Short Term	Long Term
Rock beds	Inorganic materials	Rock beds
Earth beds	Organic materials	Earth beds
Water tanks	Fatty acids	Larger water tanks
	Aromatics	Aquifers
		Solar ponds

Data from Ref. [3].

stored by converting heat. For a typical residential building, the main demands are electricity, heating, cooling, lighting, gas, hot water, and ventilation. When practical heat storage applications are considered for buildings, the heating, cooling, and hot water demands can be stored by heat storage techniques. Additionally, a power generation system like an ORC may be integrated into the system for power generation. Building demands can be stored for a short term (hourly, daily, weekly) or long term (yearly, seasonal). The selection of duration of the heat storage depends on many criteria such as storage medium and capacity. Some of the media available used in sensible and latent heat storage systems are classified in Table 2.2.

Heat storage methods are also classified according to the temperature level of the storage medium. When the temperature of the storage medium is close to environmental temperature, varying between 20 and 40°C, it is called low-temperature heat storage. When the temperature is 40–90°C, it is called as medium-temperature heat storage. If the temperature is higher than 100°C, it is called high-temperature heat storage. Also, if the heat storage is performed for cooling purposes and the temperature of the storage medium is a lower temperature than the environment, it is called cold heat storage or cold storage. If the cold heat storage performed with phase changing energy of water/ice, this heat storage technique is called an ice storage system. Details of all heat storage systems given above, are introduced in the following sections.

2.6 Sensible Heat Storage

In sensible heat storage, energy is stored by changing the temperature of a storage medium such as water, air, oil, rock beds, bricks, concretes, sand, or soil. The storage medium can consist of a single material or a few different materials. The amount of energy input to the sensible heat storage system is proportional to the difference between the final and initial temperatures of the storage medium, the mass of the storage medium, and the specific heat of storage medium. The amount of sensible heat stored in the mass of material is written as

$$Q_{\text{stored}} = mc\Delta T = \rho c_p V \Delta T \tag{2.6}$$

where c is the specific heat of the storage material, ΔT is the temperature change, ρ is the density of storage material, and m and V are the mass and volume of the storage material, respectively. It is clear in Eq. (2.6) that ΔT can be controlled with the energy charging time. To achieve higher-capacity heat storage, ΔT should be maximized. However, the density and specific heat of the storage material depends on the storage material. Each material brings its advantages and disadvantages to it. Normally, the lower storage volume is desired in a heat storage system. To achieve the lower-volume store unit, the specific heat and density of the storage material should be as high as possible. For example, the specific heat of water is approximately 4.18 kJ/kg K, and this value is approximately two times higher than the specific heat of rock and soil. The high specific heat of water often makes the heat storage in water (hot water tanks, pond, etc.) a logical choice. However, the sensible heat storage capability of water is limited by its temperature change that varies from 0 to 100°C. When the heating application is considered, the range of ΔT is between 35 and 100°C. For cooling applications, the temperature difference of the water will become 0−20°C. Although rock and soil have a lower specific heat than water, they provide higher ΔT values. Many criteria must be taken into consideration to select the optimum storage medium. The following criteria are generally considered to choose the storage medium:

- The heat storage capacity
- The thermal diffusivity of the storage material
- The storage volume restrictions due to physical conditions
- The cost of storage material
- The corrosion effects
- The lifetime of the storage material
- The type of the container keeps storage material
- The applicability to an existing system

Some common sensible heat storage materials and their properties are presented in Table 2.3. To be useful in heat storage applications, the material normally must be inexpensive and have a good specific heat. Another significant parameter for a sensible heat storage is the rate at which heat can be released and extracted. This characteristic is a function of thermal diffusivity. Iron is an excellent thermal storage medium, having both high heat capacity and high thermal conductivity. Rock is a good sensible heat storage material from the standpoint of cost, but its volumetric heat capacity is only half that of water.

A typical sensible heat storage system consists of a storage material, a container, and input/output devices. Containers should be well insulated to prevent heat losses. The common sensible heat storage systems used in practical applications can be listed as follows:

- Thermally stratified heat storage tanks
- Rock and water/rock beds heat storage
- Borehole storage
- Aquifer heat storage
- Solar ponds
- Concrete heat storage

Table 2.3 The Sensible Heat Storage Features at 20°C of Some Common Materials.

Material	Density (kg/m³)	Specific Heat (J/kg K)	Thermal Conductivity (W/kgK)	Volumetric Thermal Capacity (10^6 J/m³K)
Aluminum	2710	896		2.43
Brick	1800	837	1.3	1.51
Clay	1458	879		1.28
Concrete	2000	880	1.5	1.76
Glass	2710	837		2.27
Gravelly earth	2050	1840		3.77
Iron	7900	452		3.57
Magnetite	5177	752	5.0	3.89
Sandstone	2200	712	1.0	1.57
Steel	7840	465		3.68
Water	988	4182		4.17
Wood	700	837		1.51

Adapted from Ref. [3].

Each sensible heat storage technique listed before is very critical for meeting energy demands effectively. Therefore, they have been taken place in engineering applications for many years and investigated by scientists and technology developers. Details of these systems will be discussed in the following sections.

2.6.1 Thermally Stratified Heat Storage Tanks

Heat storage tanks are one of the oldest and common heat storage techniques. They are commonly used for domestic hot water needs, space heating, and AC applications for many years. Especially they have an important place in solar thermal applications used for hot water and space heating. Due to their relatively low initial cost, generally long lifetime, and easy-to-set storage capacity, they are used in many sectors. Additionally, they can be used in individual users (lower storage capacity) and industrial applications (higher storage capacity).

In a liquid heat storage tank such as a hot water tank, the most significant performance criteria are the thermal stratification, also called temperature stratification. Many studies related to thermal stratification and its applications emphasize that enhanced thermal stratification provides higher heat storage performance. An enhanced thermal stratification also helps

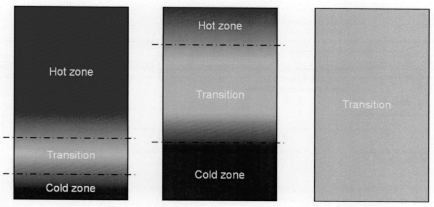

FIG. 2.9

Heat storage tanks have different degrees of thermal stratification: (left) higher thermal stratification, (middle) lower thermal stratification, and (right) no thermal stratification.

- increase the efficiencies of the system and system equipment;
- decrease the capacities and power requirement of the devices used in the system; and
- increase the flow rate of natural circulation floe depending on the density change in working fluid.

Therefore, it is always desired to keep in as higher degree as possible. Thermal stratification can be defined basically as the temperature gradient from the bottom to the top of the heat storage tank. Fig. 2.9 illustrates the different degrees of thermal stratification. In a thermally stratified heat storage tank, the tank has generally three sections: cold, transition, and hot zones. To achieve higher thermal stratification degree, the cold and transition zones should be minimum thickness. Many methods, tank designs and devices have been presented in the literature. The main of these methods is to get lower the thicknesses of the cold and transition zones. It is well known that a higher height-to-diameter ratio always provides higher thermal stratification. Also, cold water inlet feed should be done at the bottom of the tank with a lower flow rate, to prevent the mixing inside the tank due to the inlet water jet. There is no thermal stratification inside the tank in case the tank is fully mixed and the temperature is almost identical in the tank.

Since thermal stratification is very important for heat storage tanks, parallel to methods to enhance it, many methods have been presented in the literature to determine the degree of the thermal stratification. The different methods to characterize the thermal stratification are shown in Fig. 2.10.

Energy and exergy efficiencies in the thermally stratified heat storage tanks are convenient methods to determine the degree of the thermal stratification beside the methods given in Fig. 2.10. Thermodynamic analysis based on energy and exergy

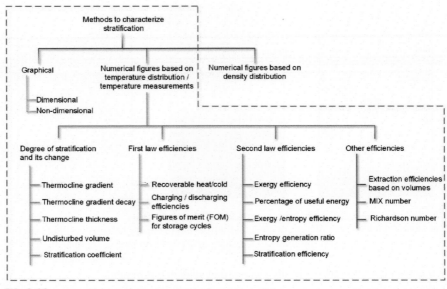

FIG. 2.10

Different methods to characterize the thermal stratification.

Adapted from Ref. [4].

calculations is performed by using the temperature gradient in the heat storage tank. The temperature gradient in the heat storage tank can be obtained by theoretical, numerical, and experimental works. In the vertical heat storage tank, it is a common way to use the temperature gradient, which is in the centerline from tank bottom to top. Details and analysis of the thermally stratified heat storage tanks are given in the case studies.

2.6.2 Rock Bed Heat Storage

Although its volumetric heat storage capacity is lower than that of water, rock is a convenient material for heat storage due to its lower cost. The biggest advantage of the rock is to able to use at a temperature higher than 100 C. Heat storage in rock bed or rock bins is generally done for space heating. They are used for storing solar energy for the short term and long term. They can be integrated into heat pump systems to improve the efficiency of the system and heat recovery. The schematic view of a solar-driven heat storage system using rock bed as a storage medium is shown in Fig. 2.11. The working principle of the system is same as other solar energy systems.

1. Charging period: The heated air or water in the solar collectors is pumped or blew to the rock bed. The temperature of the rock particle increases. This cycle is run during the daytime. If the heat storage is to be performed for long-term purposes, the charging period is operated all summer seasons.

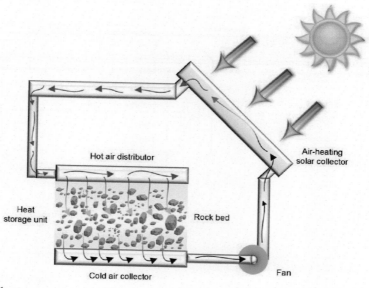

FIG. 2.11

Schematic view of the solar-driven rock bed heat system used for space heating.

2. Storing period: Rock particles with increased internal energy are kept in a bed. During the storing period, the most critical issue is the heat losses. To minimize the heat losses, either the rock bed should be insulated, or the storing period should be reduced.

3. Discharging period: The cold air or water is sent to rock bed. The temperature of working fluid increases at the outlet of the storage zone. This high-temperature working fluid can now be used for heating purposes. It is also used for pre-heating of working fluids used in energy systems such are Rankine cycle, Brayton cycle, and so on, or combustion air.

Note that caverns and abandoned mines in many paces are suitable spaces for heat storage in the rock beds. In some applications, a rock bed can be combined with water to increase heat storage capacity and thermal conductivity. However, a high capacity pump is required to pump the water in the rock bed since the rock bed is a porous medium. The pressure losses in the rock bed will be high. Therefore, in practical applications, the air is used as the working fluid, especially in long-term and high-capacity systems.

2.6.3 Borehole Heat Storage

Borehole heat storage systems are used for storing solar energy for seasonal use. It is also an example of underground heat storage techniques. It acts as a giant underground heat exchanger. A borehole heat storage system consists of many boreholes. To build a borehole heat storage system, first, the boreholes are drilled. Then, pipes

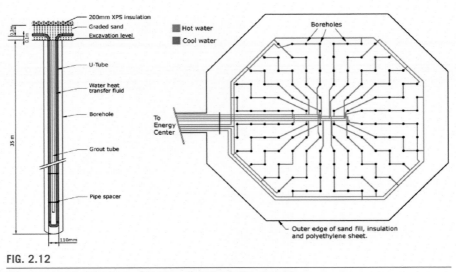

FIG. 2.12

The aerial and side views of the borehole heat storage system.

Adapted from Ref. [5].

with a U bend are inserted into the boreholes. Surrounding of the pipes is filled with materials that have high thermal conductivity. Fig. 2.12 demonstrates the borehole heat storage system in the Drake Landing Solar Community [5].

This borehole system consists of 144 boreholes, which are 37 m of depth. There is 2.25 m of the distance between pipes. The borehole heat storage facility is covered with an insulation layer and soil. Also, a landscaped park is built on top of it. During the summer season, the heated water in solar collectors is pumped to the boreholes, and heat is transferred to the surrounding soils and rock beds. The energy charging period is operated during the summer season. When the heat is needed to home, cooler water is pumped to the boreholes to heat. The heated water is sent to the short-term storage tank and then circulated to the homes through the district heating loop. Heat discharging is performed by starting from outer boreholes. In the first few years, the borehole heat storage system was not charged completely, and the storage energy was not enough to meet the heating demand of the homes. After a few years, the temperature of the borehole heat storage facility has reached almost 80 C. This temperature level is enough for meeting the heating demand of the district.

The second interesting application of the borehole heat storage system is from Ontario Tech University in Oshawa, Ontario, Canada. In the center of the campus, there are 384 boreholes, which have 213 m of depth. This system has been completed in November 2003. The facility is covered with grass for campus social activities.

2.6.4 Aquifer Heat Storage

Aquifer heat storage is an underground heat storage method used for storing large quantities of heat. Fresh water sources are generally used as the aquifer. Aquifers often have large volumes, more than millions of cubic meters. Besides water, there are also clay, sand, or rock in the aquifer. It is possible to store cold and heat capacities for daily, weekly, or seasonal in an aquifer. Fig. 2.13 illustrates the schematic view of an aquifer heat storage system used for storing heat and cooling capacities for the long term. Charging and discharging of heat is achieved by extraction and injection of groundwater from aquifers using groundwater wells. For cooling applications, the cold water in the aquifer is extracted and pumped to buildings. Then, the heated returning water is injected into the aquifer, and this high temperature of returning water creates storage of heated groundwater for heating applications. In winter, the reverse flow direction is used in the system. Stored heat capacity is used for meeting the heating demand of buildings. The heating application with an aquifer is often integrated into a heat pump system. Thus, both cooling and heating demand can be stored seasonally with an aquifer heat storage system.

Aquifer heat storage systems can be integrated renewable energy sources. Solar energy can be used for charging energy during summer for winter use. Geothermal

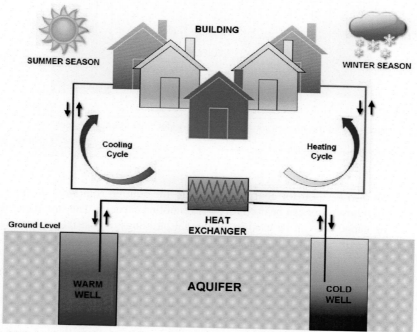

FIG. 2.13

Schematic view of an aquifer heat storage system.

Modified from Ref. [6].

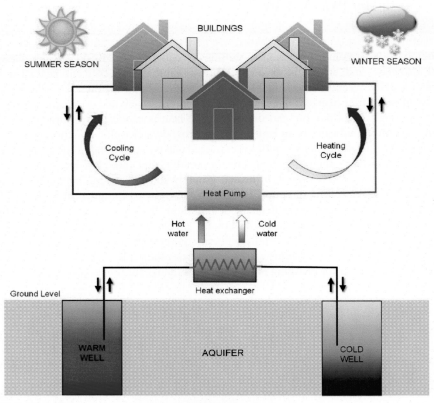

FIG. 2.14

The schematic view of the heat pump—integrated aquifer heat storage system.

Modified from Ref. [6].

sources can also be used for charging energy to the aquifer. Similarly, during the winter, cold return water can be used for storing cooling capacity. The stored cooling and heating capacities can be used in residential buildings, industrial applications, greenhouse heating, and so on. Since the energy storage capacity of the aquifer is high and long term, they can be used in many sectors.

It is very common to use aquifer heat storage systems with heat pumps. A schematic view of the heat pump—integrated aquifer heat storage system is given in Fig. 2.14. In this system, during the summer, underground water is sent from cold well to the hot well. Underground water is used in the condenser of the heat pump system. In winter, the underground water is pumped from the hot well to cold well. The heat carried from the aquifer is used in the evaporator of the heat pump.

2.6.5 Solar Ponds

Solar ponds that consist of saltwater are a heat storage method to store solar energy. Solar ponds can be used for short- and long-term heat storage purposes. A schematic view of the solar pond is shown in Fig. 2.15. Solar energy can be charged into the pond directly from the outer surface of the pond or by using solar collectors. Solar ponds that use water surface to collect solar energy are also one of the oldest heat storage techniques. In an ordinary pond or lake, when the sun's rays hit the surface of the water, while some of them reflect from the surface of the water and the remaining part are absorbed by the water, the temperature of the water increases. The heated water rises since its density decreases. When it arrives at the surface, its temperature reduces. Consequently, the final temperature of the pond is almost equal to the temperature of the atmosphere. In the solar ponds, salt is dissolved at the bottom of the pond to increase the density of the water. Thus, the water heated during the energy charging always stays at the bottom part of the solar pond. In a solar pond, the salt concentration increases with increases in depth. The useful energy located in the bottom part of the solar pond can be discharged via a heat exchanger. In advanced solar pond applications, heat exchangers and solar collectors are used for charging energy into the water. A working fluid is circulated between the heat exchanger and solar collectors.

The biggest advantage of the solar pond is its reasonable initial cost. Although it is more reasonable to use in the regions where the land structure is suitable, creating a solar pond volume today is not a very costly and time-consuming process. For this reason, in regions where solar radiation is good enough, an artificial pond or lake may be created, and heat storage can be achieved in a solar pond.

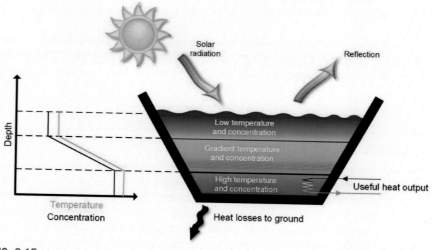

FIG. 2.15

The schematic view of the solar pond.

In the solar pond heat storage system, clear water is used for increasing the penetration of the sun's rays. Also, the bottom of the pond is generally black to increase the absorption of sun's rays. As seen from Fig. 2.15, there are three zones in a solar pond: surface, transition, and storage zones. In the surface zone, there is fresh water, and the water temperature is almost equal to the surrounding temperature. The transition zone is a layer between the salty storage zone and surface zone. The transition zone acts as an insulation layer since it prevents the natural convention between storage and surface zones. The thickness of the transition is important since it directly affects the volume of the effective storage zone. The stored heat in the solar pond may be used for heating purposes such as space heating, power generation with an ORC, and hot water.

2.6.6 Concrete Heat Storage

Concrete is used for heat storage in buildings for both heating and cooling applications. It is an important storage medium due to its low cost and availability in all buildings. Concrete has also that following characteristics as a heat storage medium [3]:

- high specific heat,
- good mechanical strength,
- lower thermal expansion coefficient, and
- high mechanical resistance to cycle energy charging and discharging.

The concrete columns can be used for heat storage purposes since they exist in a great volume in the buildings. Ozrahat and Unalan [7] have comprehensively investigated the use of concrete columns for sensible heat storage and space heating applications. They have built a concrete column that has a square cross section. They have concluded that concrete column can be used as sensible heat storage and space heating.

In a sensible heat storage systems, which consist of a concrete columns, waste heat from the building or solar energy can be used as an energy source. The stored energy can be discharged by air flowing through the concrete columns (Fig. 2.16).

2.7 Latent Heat Storage

Latent heat storage is performed based on phase changing of a substance. The biggest advantage of the latent heat storage is having a higher energy storage capacity than sensible heat storage for a given substance. As an example, sensible and latent heat change for water is discussed in Section 2.3. Briefly, when phase change for 1 kg of water under atmospheric pressure is considered from ice at $-10°C$ to vapor at $100°C$, the sensible heat change is approximately 428.5 kJ while the latent heat change is 2593 kJ. Therefore, latent heat storage methods have a big potential to store higher capacities in a lower storage volume. Enthalpy changing for water is

FIG. 2.16

The experimental setup for sensible heat storage system using a concrete column.

Adapted from Ref. [7].

also illustrated in Fig. 2.17. This enthalpy changing trend is similar to all pure substances. In Fig. 2.17, point A denotes −10°C of ice. From A to B, the ice absorbs heat and its temperature increases up to 0°C due to sensible heat change. When the temperature of ice reaches the 0°C, it starts melting (latent heat change). During melting, the temperature of ice—water mixture stays constant at the temperature of 0°C. For melting of 1 kg ice, it is required to charge 333.6 kJ of energy (from B to C). At the point C, there is liquid water at the temperature of 0°C. The temperature of water increases with heat input (from C to D). When the temperature of water reaches the 100°C, it starts to boil, and the boiling process occurs until all liquid water turns the vapor. The temperature and enthalpy values for water seen in Fig. 2.17 are summarized in Table 2.4.

One of the critical issues in latent heat storage methods is the volume changing during the phase change process. The volume change in the phase changing (expansion or shrink) is generally higher than sensible heat change. Therefore, latent heat storage systems and storage mediums require more complicated and strength storage containers than sensible heat storage systems. This of course affects the cost of the system considerably.

In latent heat storage systems, during heating and cooling processes, the storage medium undergoes a phase change. This phase change for heat storage applications is generally solidification and melting. Gasification and liquefaction are generally not preferred in heat storage applications because of higher volume change.

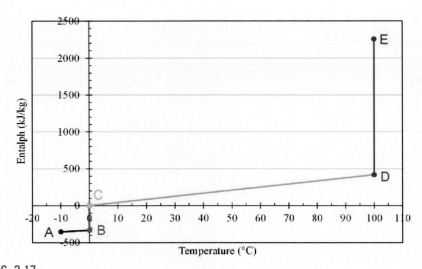

FIG. 2.17

Enthalpy change of water during phase changing.

Table 2.4 Temperature and Enthalpy Values for the State Points Given.

Point	Temperature (°C)	State	Enthalpy (kJ/kg)
A	−10	Ice (solid)	−354.1
B	0	Ice (solid)	−333.1
C	0.01	Water (liquid)	0.001
D	100	Water (liquid)	419.1
E	100	Water (vapor)	2675.5

The storage medium used in latent heat storage is called PCM. Energy changing in a PCM during phase changing shown in Fig. 2.17 can be defined as

$$\Delta E_{PCM} = \int_{h_A}^{h_E} m\, dh = m(h_E - h_A)$$

$$\Delta E_{PCM} = \int_{T_A}^{T_B} mc_{solid}(T_B - T_A) + m\, h_{sf} + \int_{T_C}^{T_D} mc_{liquid}(T_D - T_C) + m\, h_{fg}$$

(2.7)

Here, h_{sf} and h_{fg} are the latent heats of fusion and vaporization of the PCM, respectively. c_{solid} and c_{fluid} are the specific heats of the PCM for solid and liquid phases. There are many PCMs and their different heat storage methods. PCMs and their advantages and disadvantages are discussed in the following section.

An ordinary latent heat storage system primarily consists of the following components:

- PCM: PCM is the medium where energy is charged, stored, and discharged. The most significant two parameters for a PCM are phase change enthalpy and its temperature. Higher phase change enthalpy provides lower storage volume for a given storage capacity. The selection of phase change temperature directly depends on the energy source. Proper PCM should be selected to design latent heat storage optimally.
- PCM container: PCMs are kept in specially designed containers to control the phase change of PCMs. PCM container should be well designed due to increase in the thermal performance of the entire system. PCMs are kept in small capsules (encapsulation) or high-volume tanks or vessels consisted of tube banks.
- Heat transfer fluid: PCMs cannot be used directly in energy conversion devices due to phase changing. Therefore, a heat transfer fluid is used in a latent heat storage system to charge energy into the PCM and discharge energy from the PCM. Thermophysical specifications are significant for the thermal performance of the system. They should be selected accordingly.

A latent heat storage system provides lower storage volume than a sensible heat storage system for a given storage capacity (Fig. 2.18). This lower volume also means a lower heat loss due to lower surface area. Thermal stress because of temperature change in the system is generally lower and more manageable in latent heat storage system, as phase change occurs in a constant temperature. However, latent heat storage systems have some disadvantages due to complexity of phase change processes. The most significant issue is the volume change during phase change. Latent heat storage systems should be designed by considering this much of volume changing. Therefore, sublimation and deposition processes are generally not used in latent heat storage system as they cause high volume changing.

2.7.1 Phase Change Materials

When a material melts, vaporizes, or sublimates, it absorbs heat. In the reverse processes that solidify, condensation or deposition, it releases heat. An illustration of phase changes of a material is given in Fig. 2.19. The required or released energies

FIG. 2.18

Sensible and latent heat storage capacities of 1 kg of water.

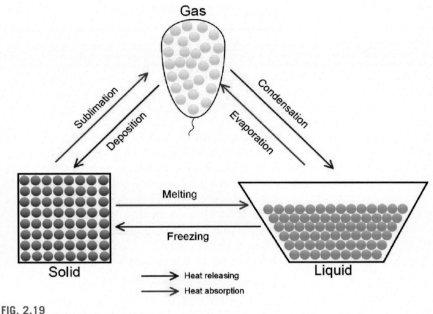

FIG. 2.19

Illustration of phase changes for a material.

can be used for heat storage purposes. The material changing its phase during thermal processes is called PCM. PCMs are not only used in heat storage purposes, but they are also used for thermal management purposes. For heat storage applications, solid−liquid PCMs are generally used due to the amount of volume change. When the heat storage capacities and the amount of PCMs are taken into consideration, it is hard to handle and manage volume changes during phase changing for liquid−gas and solid−gas PCMs. Ice/water, eutectic salts, and molten salts are used since ancient times as PCMs for heat storage purposes. One of the oldest PCM applications was seat heaters used in British trains in the late 1800s. Sodium thiosulfate pentahydrate was used as the PCM, which melts and solidifies at 44.4°C of temperature. The use of PCMs has increased day by day since the late 1800s with increasing energy demand and the need for effective thermal management. Today, PCMs are a unique solution for both heat storage and thermal management applications.

The PCMs used in latent heat storage systems are generally classified as organics, inorganics, and eutectics. Besides, there are miscellaneous PCMs under development. A common classification of PCMs used in latent heat storage applications is shown in Fig. 2.20. The PCM properties have a great impact on the performance of the latent heat storage system. The characteristics of the PCMs are listed as

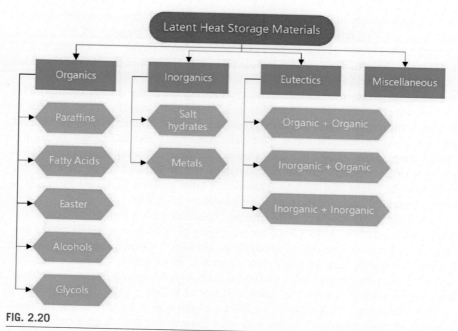

FIG. 2.20

Classification of phase change materials.

- phase change temperature range,
- latent heat of fusion,
- specific heat,
- thermal conductivity,
- density,
- thermal expansion during the phase change,
- subcooling effect,
- chemical stability,
- type of packaging,
- reversible phase change,
- toxicity,
- flammable,
- cost, and
- availability.

Each item has a significant effect on both the thermal and economic performances of the latent heat storage system. Phase change temperature is generally determined according to input and output temperatures. It should higher than output temperature and lower than the input temperature to provide fully melting and solidifying. The latent heat of PCM is a decisive effect on the volume of storage unit. Higher latent heat provides lower storage unit volume for a given storage capacity. PCM density has also the effect of the volume of storage units. Lower PCM density provides a lower storage volume for a given storage capacity and fusion enthalpy.

In some applications, the lower storage mass may be significant due to the physical conditions of the environment. In this case, the lower-density PCMs can be preferred. For example, if the storage unit is to be placed on the roof, a lower weight is generally desired. This plays a critical role on the statics of the building.

The specific heat and thermal conductivity of PCMs for both phases have a great impact on the progress of solidification and melting inside the storage medium. The determination of thermal conductivity is a significant issue for PCMs. There is limited information in the literature about the thermal conductivity of PCMs and the determination method for it. Thermal conductivity and its change during the phase changing are studied in recent studies. The determination thermal conductivity of the PCMs is very critical to understand the progress of phase changing for analyzing the latent heat storage systems and other systems that used PCMs. It is also an essential issue for the numerical modeling of PCMs.

The thermal expansion of PCMs during phase change is a significant issue to the design and cost of the system since the packaging of PCM is required. It causes mechanical stress in the packaging container in addition to thermal stress. To solve the problem caused by thermal expansion, generally, an air gap is left inside the containers such as the capsule, tube, and so on. Although the air gap decreases the thermal performance, it absorbs the volume change of PCM.

The subcooling effect is basically defined as the start to solidify below the temperature of phase change. It is the undesired issue as it causes higher energy consumption. To solve this issue, generally, some additives are added inside the PCM. It is also a significant issue to start to crystallize from the center of PCM to improve the thermal performance of PCM because the thermal conductivity of the solid phase of PCM is generally lower than the liquid phase. If the crystallization starts from the edge of the PCM, the solid PCM layer causes a thermal resistance. Therefore, some additives are added to PCM to solve this issue.

The type of packaging of PCMs is an essential matter for both performance and cost of the system. Capsules and tube banks are widely used for packaging in practical applications. In addition to these parameters, toxicity, flammable, availability, and cost features of the PCMs are significant for systems that used PCMs. In heat storage systems, it is expected to use nontoxicity and nonflammable PCMs. The availability and cost are significant parameters for the practical application of PCMs. It is critical to use abundant and low-cost PCMs to improve the economic performance of the system.

There are many PCMs used in research projects and practical applications. In Table 2.5, some organic, inorganic, ester and fatty acid, alcohol and glycol PCMs are listed. These PCMs are used in many heat storage applications. As mentioned earlier, the selection of PCMs depends on many criteria such as latent heat capacity, thermal conductivity, melting point, density, abundance, cost, and so on.

In addition to PCMs listed in Table 2.5, kinds of paraffin, also called paraffin waxes, are the members of the hydrocarbon family and widely used in latent heat storage applications. They are produced in petroleum distillation processes as a side product. There is fatty content in it, and their thermophysical features depend

Table 2.5 The List of Some PCMs for Organic, Inorganic, and Identified Number of Esters, Fatty Acids, Alcohols, and Glycols Used in Latent Heat Storage Applications.

PCM	Melting Point (°C)	Latent Heat (kJ/kg)	PCM	Melting Point (°C)	Latent Heat (kJ/kg)
Organic PCMs					
N-tetradecane	5.5	226	P-dichlorobenzene	53.1	121
Formic acid	7.8	247	N-pentacosane	53.7	164
N-pentadecane	10.0	205	Myristic acid	54.0	199
Acetic acid	16.7	273	Oxolate	54.3	178
N-hexadecane	16.7	237	Tristearin	54.5	191
Caprilone	40.0	260	O-xylene dichloride	55.0	121
Docosyl bromide	40.0	201	β-chloroacetic acid	56.0	147
N-henicosane	40.5	161	N-hexacosane	56.3	255
Phenol	41.0	120	Nitro naphthalene	56.7	103
N-lauric acid	43.0	183	α-chloroacetic acid	61.2	130
P-joluidine	43.3	167	N-octacosane	61.4	134
Cynamide	44.0	209	Palmitic acid	61.8	164
N-docosane	44.5	157	Bees wax	61.8	177
N-tricosane	47.6	130	Glycolic acid	63.0	109
Hydrocinnamic acid	48.0	118	P-bromophenol	63.5	86
Cetyl alcohol	49.3	141	Azobenzene	67.1	121
O-nitroaniline	50.0	93	Acrylic acid	68.0	115
Camphene	50.0	239	Dinitrotoluene (2,4)	70.0	111
Diphenyl amine	52.9	107	Phenylacetic acid	76.7	102
			Thiosinamine	77.0	140
			Benzylamine	78.0	174

Continued

Table 2.5 The List of Some PCMs for Organic, Inorganic, and Identified Number of Esters, Fatty Acids, Alcohols, and Glycols Used in Latent Heat Storage Applications.—cont'd

PCM	Melting Point (°C)	Latent Heat (kJ/kg)	PCM	Melting Point (°C)	Latent Heat (kJ/kg)
Inorganic PCMs					
H_2O	0.0	333	$SO_3(\beta)$	32.3	151
$POCl_3$	1.0	85	$TiBr_4$	38.2	23
D_2O	3.7	318	$H_4P_2O_6$	55.0	213
$SbCl_5$	4.0	33	$SO_3\ (\gamma)$	62.1	331
H_2SO_4	10.4	100	$SbCl_3$	73.4	25
IC 1 (β)	13.9	56	$NaNO_3$	307	199
MOF_6	17.0	50	KNO_3	380	266
$SO3(\alpha)$	17.0	108	KOH	380	149
IC 1 (α)	17.2	69	$MgCl_2$	800	492
P_4O_6	23.7	64	NaCl	802	492
H_3PO_4	26.0	147	Na_2CO_3	854	275
Cs	28.3	15	KF	857	452
Ga	30.0	80	K_2CO_3	897	235
$AsBr_3$	30.0	38	$SnBr_4$	30.0	28
Esters, Fatty acids, Alcohols, and Glycols					
Formic acid	7.8	247	Acetamide	81	241
Acetic acid	16.7	187	Methyl bromobenzoate	81	126
Glycerin	17.9	198.7	α-Naphthol	96	163
Lithium chloride ethanolate	21	188	Glutaric acid	97.5	156
Polyethylene glycol 600	20–25	146	p-Xylene dichloride	100	138.7

D–Lactic acid	26		Methyl fumarate	102
1-3 methyl pentacosane	29		Catechol	104.3
Camphenilone	39		Quinone	115
Docosyl bromide	40		Acetanilide	115
Caprylone	40		Succinic anhydride	119
Heptadecanone	41		Benzoic acid	121.7
1-Cyclohexyloctadecane	41		Stibene	124
4-Heptadacanone	41		Benzamide	127.2
Cyanamide	44		Phenacetin	137
Methyl eicosanoate	45		α-Glucose	141
3-Heptadecanone	48		Acetyl-p-toluidene	146
2-Heptadecanone	48		Durene	79.3
Camphene	50		Salicylic acid	159
9-Heptadecanone	51		Benzanilide	161
Methyl behenate	52		O-mannitol	166
Pentadecanoic acid	52.5		Hydroquinone	172.4
Hypophosphoric acid	55		p-Aminobenzoic acid	187
Chloroacetic acid	56		Heptadecanoic acid	60.6
Bee wax	61.8		Bromcamphor	77
Glycolic acid	63		Arachic acid	76.5
Trimyristin	33–57		Phenylhydrazone benzaldehyde	155

	184			242
	197			207
	205			171
	201			142
	259			204
	201			142.8
	218			167
	197			169.4
	209			136.7
	230			174
	218			180
	218			156
	238			199
	213			162
	234			294
	178			258
	213			153
	130			189
	177			174
	109			227
	201–213			134.8

Modified from Ref. [8].

Table 2.6 The List of Paraffins Used in Latent Heat Storage Applications.

Name	Number of Carbon Atoms	Melting Point (°C)	Density (kg/m³)	Thermal Conductivity (W/mK)	Latent Heat (kJ/kg)
n-Dodecane	12	−12	750	0.21^S	n.a.
n-Tridecane	13	−6	756		n.a.
n-Tetradecane	14	4.5−5.6	771		231
n-Pentadecane	15	10	768	0.17	207
n-Hexadecane	16	18.2	774	0.21^S	238
n-Heptadecane	17	22	778		215
n-Octadecane	18	28.2	$814^S,775^L$	$0.35^S, 0.149^L$	245
n-Nonadecane	19	31.9	$912^S,769^L$	0.21^S	222
n-Eicosane	20	37			247
n-Heneicosane	21	41			215
n-Docosane	22	44			249
n-Tricosane	23	47			234
n-Tetracosane	24	51			255
n-Pentacosane	25	54			238
Paraffin wax	n.a.	32	$785^S,749^L$	$0.514^S,0.224^L$	251
n-Hexacosane	26	56	770	0.21^S	257
n-Heptacosane	27	59	773		236
n-Octacosane	28	61	$910^S,765^L$		255
n-Nonacosane	29	64			240
n-Triacontane	30	65			252
n-Hentriacontane	31	n.a.	$930^S,830^L$		n.a.
n-Dotriacontane	32	70			n.a.
n-Tritriacontane	33	71			189

Modified from Ref. [8].

on the ratio of fat. The kinds of paraffin used in the latent heat storage application are listed in Table 2.6.

Note that salt hydrates and eutectics are recognized as one of the oldest and widely used PCMs in heat storage applications. Salt hydrates consist of a salt and water. They can combine with eutectics. The melting points of salt hydrates vary from 15 to 120°C. Salt hydrates have advantages because of their low cost and abundance. Due to salt contents, they have higher thermal conductivity. Their latent heat capacity is higher than many materials used in latent heat storage applications. One of the most significant advantages of salt hydrates is the low volume change during phase changes. However, the packaging material for salt hydrates is a critical issue due to their salt content. Some salt hydrates are listed in Table 2.7. Eutectics consist of two or more components that melt and solidify at the same temperature. They can be used with salt hydrates to increase heat storage density. Some eutectics are listed in Table 2.8.

Table 2.7 The List of Salt Hydrates Used in Latent Heat Storage Applications.

Name	Melting Point (°C)	Density (kg/m³)	Thermal Conductivity (W/mK)	Latent Heat (kJ/kg)
$LiClO_3 \cdot 3H_2O$	8	n.a.	n.a.	253
$NH_4Cl \cdot Na_2SO_4 \cdot 10H_2O$	11	n.a.	n.a.	163
$K_2HO_4 \cdot 6H_2O$	14	n.a.	n.a.	108
$NaCl \cdot Na_2SO_4 \cdot 10H_2O$	18	n.a.	n.a.	286
$KF \cdot 4H_2O$	18	n.a.	n.a.	330
$K_2HO_4 \cdot 4H_2O$	18.5	1447^{20C}, 1455^{18C}, 1480^{6C}	n.a.	231
$Mn(NO_3)_2 \cdot 6H_2O$	25 / 25.8	1738^{20C}, 1728^{40C}, 1795^{5C}	n.a.	148 / 125.9
$LiBO_2 \cdot 8H_2O$	25.7	n.a.	n.a.	289
$FeBr_3 \cdot 6H_2O$	27	n.a.	n.a.	105
$CaCl_2 \cdot 6H_2O$	29–30	1562^{32C}, 1802^{24C}	$0.561^{61.2C}$, 1.008^{23C}	170–192
$LiNO_3 \cdot 3H_2O$	30	n.a.	n.a.	189–296
$Na_2SO_4 \cdot 10H_2O$	32	1485^{24C}	0.544	251–254
$Na2CO_3 \cdot 10H_2O$	33–36	1442	n.a.	247
$KFe(SO_4)_2 \cdot 12H_2O$	33	n.a.	n.a.	173
$CaBr_2 \cdot 6H_2O$	34	1956^{35C}, 2194^{24C}	n.a.	115–138
$LiBr \cdot 2H_2O$	34	n.a.	n.a.	124
$Na_2HPO_4 \cdot 12H_2O$	35	1522	n.a.	256–281
$Zn(NO_3)_2 \cdot 6H_2O$	36	1828^{36C}, 1937^{24C}, 2065^{14C}	$0.464^{39.9C}$, $0.469^{61.2C}$	134–147
$Mn(NO_3)_2 \cdot 4H_2O$	37	n.a.	n.a.	115
$FeCl_3 \cdot 6H_2O$	37	n.a.	n.a.	223
$CaCl_2 \cdot 4H_2O$	39	n.a.	n.a.	158

Continued

Table 2.7 The List of Salt Hydrates Used in Latent Heat Storage Applications.—*cont'd*

Name	Melting Point (°C)	Density (kg/m³)	Thermal Conductivity (W/mK)	Latent Heat (kJ/kg)
$CoSO_4 \cdot 7H_2O$	40.7	n.a.	n.a.	170
$CuSO_4 \cdot 7H_2O$	40.7	n.a.	n.a.	171
$KF \cdot 2H_2O$	42	n.a.	n.a.	162–266
$MgI_2 \cdot 8H_2O$	42	n.a.	n.a.	133
$CaI_2 \cdot 6H_2O$	42	n.a.	n.a.	162
$Ca(NO_3)_2 \cdot 4H_2O$	43–47	n.a.	n.a.	106–140
$Zn(NO_3)_2 \cdot 4H_2O$	45	n.a.	n.a.	110
$K_3PO_4 \cdot 7H_2O$	45	n.a.	n.a.	145
$Fe(NO_3)_3 \cdot 9H_2O$	47	n.a.	n.a.	155–190
$Mg(NO_3)_3 \cdot 4H_2O$	47	n.a.	n.a.	142
$Na_2SiO_3 \cdot 5H_2O$	48	n.a.	n.a.	168
$Na_2HPO_4 \cdot 7H_2O$	48	n.a.	n.a.	135–170
$Na_2S_2O_3 \cdot 5H_2O$	48	1600	n.a.	209
$K_2HPO_4 \cdot 3H_2O$	48	n.a.	n.a.	99
$MgSO_4 \cdot 7H_2O$	48.4	n.a.	n.a.	202
$Ca(NO_3)2 \cdot 3H_2O$	51	n.a.	n.a.	104
$Na(NO_3)_2 \cdot 6H_2O$	53	n.a.	n.a.	158
$Zn(NO_3)_2 \cdot 2H_2O$	55	n.a.	n.a.	68
$FeCl_3 \cdot 2H_2O$	56	n.a.	n.a.	90
$CO(NO_3)_2 \cdot 6H_2O$	57	n.a.	n.a.	115
$Ni(NO_3)_2 \cdot 6H_2O$	57	n.a.	n.a.	168
$MnCl_2 \cdot 4H_2O$	58	n.a.	n.a.	151
$CH_3COONa \cdot 3H_2O$	58	n.a.	n.a.	270–290
$LiC_2H_3O_2 \cdot 2H_2O$	58	n.a.	n.a.	251–377
$MgCl_2 \cdot 4H_2O$	58	n.a.	n.a.	178
$NaOH \cdot H_2O$	58	n.a.	n.a.	272
$Na(CH_3COO) \cdot 3H_2O$	58	n.a.	n.a.	n.a.
$Cd(NO_3)_2 \cdot 4H_2O$	59	n.a.	n.a.	98
$Cd(NO_3)_2 \cdot 1H_2O$	59.5	n.a.	n.a.	107
$Fe(NO_3)_2 \cdot 6H_2O$	60	n.a.	n.a.	125
$NaAl(SO_4)_2 \cdot 12H_2O$	61	n.a.	n.a.	181
$FeSO_4 \cdot 7H_2O$	64	n.a.	n.a.	200
$Na_3PO_4 \cdot 12H_2O$	65	n.a.	n.a.	168
$Na_2B_4O_7 \cdot 10H_2O$	68	n.a.	n.a.	n.a.
$Na_3PO_4 \cdot 12H_2O$	69	n.a.	n.a.	n.a.
$LiCH_3COO \cdot 2H_2O$	70	n.a.	n.a.	150–251
$Na_2P_2O_7 \cdot 10H_2O$	70	n.a.	n.a.	186–230
$Al(NO_3)_2 \cdot 9H_2O$	72	n.a.	n.a.	155–176

Table 2.7 The List of Salt Hydrates Used in Latent Heat Storage Applications.—*cont'd*

Name	Melting Point (°C)	Density (kg/m³)	Thermal Conductivity (W/mK)	Latent Heat (kJ/kg)
$Ba(OH)_2 \cdot 8H_2O$	78	1937^{84C}, 2070^{24C}, 2180	$0.653^{85.7C}$, $0.678^{98.2C}$, 1.255^{23C}	265–280
$Al_2(SO_4)_3 \cdot 18H_2O$	88	n.a.	n.a.	218
$Sr(OH)_2 \cdot 8H_2O$	89	n.a.	n.a.	370
$Mg(NO_3)_2 \cdot 6H_2O$	89–90	1550^{94C}, 1636^{25C}	0.490^{95C}, 0.502^{110C}, 0.611^{37C}, $0.669^{55.6C}$	162–167
$KAl(SO_4)_2 \cdot 12H_2O$	91	n.a.	n.a.	184
$(NH_4)Al(SO_4) \cdot 6H_2O$	95	n.a.	n.a.	269
$Na_2S \cdot 51/2H_2O$	97.5	n.a.	n.a.	n.a.
$LiCl \cdot H_2O$	99	n.a.	n.a.	212
$CaBr_2 \cdot 4H_2O$	110	n.a.	n.a.	n.a.
$Al_2(SO_4)_2 \cdot 16H_2O$	112	n.a.	n.a.	n.a.
$MgCl_2 \cdot 6H_2O$	115–117	1450^{120C}, 1442^{78C}, 1569^{20C}, 1570^{20C}	0.570^{120C}, 0.598^{140C}, $0.694^{9 0C}$, 0.704^{110C}	165–169
$NaC_2H_3O_2 \cdot 3H_2O$	137	1450	n.a.	172

Modified from Ref. [8].

2.7.1.1 Hybrid (sensible + latent) heat storage

Sensible and latent heat storage mediums are sometimes combined to either increase storage capacity or control the temperature level of the storage container. In this kind of heat storage systems, there are at least two storage mediums in the storage container for sensible and latent heat storages. The most common application of hybrid heat storage systems is the PCM-integrated hot water tanks. A PCM(s) is added to a hot water tank to increase storage capacity and to control the temperature level of the storage tank. Hybrid heat storage in hot water tanks is a significant application for solar domestic hot water systems (SDHWSs). The charging period for an SDHWS is all daytime period. During the daytime, if the produced hot water is not used, the temperature of the stored hot water reaches excessive levels. Especially, in the locations that get higher solar radiations, the hot water temperature reaches more than 90°C, and it can even boil. This high temperature can damage the system and users. An illustration of PCM capsules—integrated hot water tanks is given in Fig. 2.21. Paraffin is used as the PCM, and it is placed inside the tank with

Table 2.8 The List of Some Eutectics Used in Latent Heat Storage Applications.

Name	Composition (wt%)	Melting Point (°C)	Latent Heat (kJ/kg)
Na_2SO_4 + NaCl + KCl + H_2O	31 + 13+16 + 40	4	234
Na_2SO_4 + NaCl + NH_4Cl + H_2O	32 + 14+12 + 42	11	n.a.
$C_5H_5C_6H_5$ + $(C_6H_5)_2O$	26.5 + 73.5	12	97.9
Na_2SO_4 + NaCl + H_2O	37 + 17 + 46	18	n.a.
Na_2S_4 + $MgSO_4$ + H_2O	25 + 21 + 54	24	n.a.
$C_{14}H_{28}O_2$ + $C_{10}H_{20}O_2$	34 + 66	24	147.7
$Ca(NO)_3 \cdot 4H_2O$ + $Mg(NO)_3 \cdot 6H_2O$	47 + 53	30	136
NH_2CONH_2 + NH_4NO_3	–	46	95
$Mg(NO_3)_2 \cdot 6H_2O$ + NH_4NO_3	61.5 + 38.4	52	125.5
$Mg(NO_3)_2 \cdot 6H_2O$ + $MgCl_2 \cdot 6H_2O$	58.7 + 41.3	59	132.2
$Mg(NO_3)_2 \cdot 6H_2O$ + $Al(NO_3)_2 \cdot 9H_2O$	53 + 47	61	148
$Mg(NO_3)_2 \cdot 6H_2O$ + $MgBr_2 \cdot 6H_2O$	59 + 41	66	168
Naphthalene + Benzoic acid	67.1 + 32.9	67	123.4
$AlCl_3$ + NaCl + $ZrCl_2$	79 + 17 + 4	68	234
$AlCl_3$ + NaCl + KCl	66 + 20 + 14	70	209
NH_2CONH_2 + NH_4Br	66.6 + 33.4	76	151
$LiNO_3$ + NH_4NO_3 + $NaNO_3$	25 + 65 + 10	80.5	113
$AlCl_3$ + NaCl + KCl	60 + 26 + 14	93	213
$AlCl_3$ + NaCl	66 + 34	93	201
$NaNO_2$ + $NaNO_3$ + KNO_3	40 + 7 + 53	142	n.a.

cylindrical capsules. Details of the use of PCM in the hot water, as hybrid heat storage application, are discussed in the case studies.

Due to high-temperature hot water, inattentive hot water users may be injured since in many SDHWSs, hot water is directly connected to the line. In addition to that, the pressure inside the tank increases with increasing temperature. A fair amount of heat transfer fluid used in the collector and tank loop is removed through a check valve automatically. If the descending heat transfer fluid is not added to the system, it is not to work properly, and the performance of the system decreases dramatically. Therefore, the temperature level of the tank should be controlled. PCM placement inside the tank has a significant potential to control the temperature level in the storage tank. This issue is generally seen in the SDHWSs that have vacuum tube collectors.

One of the reasons for using PCM inside the tank is to prevent the rapid cooling of the stored water during hot water consumption in the evening. During the energy discharging period, the cold-water feed by the main line destroys the thermal stratification inside the tank. Destroyed thermal stratification reduces the temperature of stored hot water immediately. PCMs can play a critical role in keeping thermal

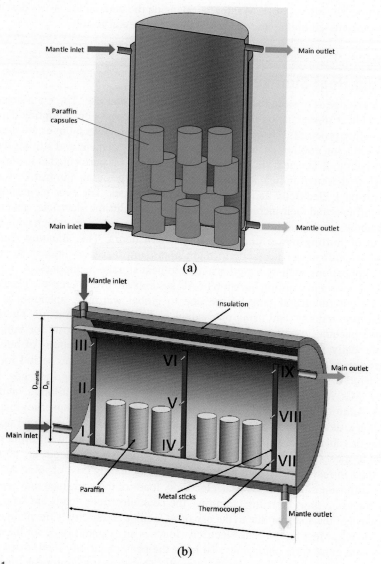

FIG. 2.21

An illustration of PCM capsules—integrated hot water tanks used in SDHWS: (a) vertical hot water tank and (b) horizontal hot water tank. *PCM*, phase change material; *SDHWS*, solar domestic hot water system.

stratification. The effects of PCM integration inside the tank on the issues described earlier are discussed in the case studies with experimental results.

2.8 Cold Heat Storage

Cold heat storage is to store the cooling capacities for later use. The main objective of the cold heat storage is to reduce cooling costs and energy consumption depending on cooling processes. However, today, cold heat storage techniques are used for not only reducing cooling costs but also grid load management and storing renewables for later use. The heat storage for cooling purposes is performed at the temperature below the environment temperature. Heat storage is an innovative way of storing off-peak electricity and renewable energy sources. The ever-increasing use of wind turbines and their increasing contribution to nighttime electricity generation brought about a rising nighttime surplus of electricity generation, which adds to the attraction of shifting power consumption to night hours. This situation makes it attractive to shift power demands to night hours and increases interest in energy storage applications such as batteries and cold heat storage techniques.

The capacity and working duration of AC systems are ever increasing because of increasing environmental temperature due to global warming, spending a more extended time indoors, and using many devices producing heat indoor. Therefore, ACs have become important operating and initial investment costs for buildings. When the daily, seasonal, and yearly peak cooling load distributions of buildings are considered, the peak cooling loads are seen in a limited time period for each period. High-capacity ACs are required to meet the peak cooling demands. High-capacity ACs increase the initial cost of HVAC (heating, ventilating, and AC) systems. Additionally, these high-capacity devices increase the capacities and costs of auxiliary devices such as grid connection transformer, power generator, the diameter of cables, and so on. Consequently, a substantial amount of initial cost is required to meet the peak cooling loads. Cold heat storage techniques are an essential method to reduce peak cooling loads and the operating cost of ACs. ACs also have a significant impact on the peak electricity loads of the grid since they are operated at the same time. In many countries, the peak electricity loads are seen in the hottest days of the summer due to the use of ACs since electricity is the main energy source of ACs. Meeting the peak electricity loads is an essential issue for the energy supplier and generation companies. To meet the peak electricity load, extra power plants are built or electricity imported directly. Therefore, meeting peak electricity loads requires a high cost. Cold heat storage systems play a critical role in shifting the peak electricity loads dependent on AC systems. Thus, both the operating cost of AC systems and the peak electricity loads are reduced by cold heat storage systems.

Cold heat storage methods can also be used for meeting the emergency cooling needs. In buildings where cooling is very critical such as supercomputer centers, data-processing centers, and stem cell centers, there are a few backup chillers, and generators used in power cutoff. Also, today, we have a critical issue with AC systems. During the COVID-19 pandemic, we have seen that almost all intensive care units and hospitals are working with overcapacity conditions. These overcapacity working conditions have caused poorly meeting the clean air and AC demands. Additionally, the capacity increases in these systems are required due to converting regular patient services into an intensive care unit. Besides, fresh air need has become a critical issue for the buildings, which have high people circulation such as malls, hypermarkets, and so on, especially in current extraordinary days. These needs require the capacity increasing in the AC systems. Cold heat storage systems are a unique solution to meet the increase in cooling capacities. While the cold heat storage systems are used for reducing cooling costs during the normal days, they can be used for meeting emergency cooling capacities in tough days such as a pandemic and long-duration power cutoff.

2.8.1 Working Principles of Cold Heat Storage Systems

Cold heat storage is performed by cooling or solidifying the storage medium. The aim of the cold heat storage is to reduce cooling costs and the capacities of the system's equipment and to shift peak cooling loads. To provide these aims, cold heat storage systems consist of three basic steps like in every energy storage system (Fig. 2.8).

Charging period: During the electricity off-peak hours, the temperature of the storage medium is either reduced or solidified. If the aim of storage is to store renewables, the cold medium is produced when renewable energy sources are active and lower demand periods.

Storing period: The produced cold or solid storage medium is kept in an insulated storage tank or vessel. During the storing period, the heat gains from the surrounding should be minimized.

Discharging period: The stored cold medium is used during the electricity peak periods. The cooling demand is met by using the stored energy. Thus, the cooling is performed by a lower cost or reduced peak demands.

These periods are usually distributed throughout the day if the storage is not to perform seasonally. They are generally planned according to electricity tariff periods. Different tariff structures are used for promoting energy conservation and reduce peak demand for electricity. Typically, the day is divided into three periods: midpeak period, peak period, and off-peak period. Depending on the energy management strategies of countries and electricity supplier companies, the number of periods may increase. However, the off-peak period always coincides with the night hours in which the energy consumption is lower. In the off-peak period, the electricity unit prices are the lowest. Cold heat storage systems reduce the operating cost of AC systems by benefiting electricity tariffs. In a typical cold heat storage system, the cold storage medium is produced during the off-peak hours (with lowest electricity unit price), and cooling carries out by using the stored cold medium when the midpeak and/or peak periods. Thus, the operating cost of AC may be decreased significantly.

2.8.2 Benefits of Cold Heat Storage Systems

Cold energy storage systems mainly provide two significant advantages. The first is to reduce the energy costs and the second is to reduce the peak cooling load of the buildings. Reducing energy costs is carried out thanks to electricity tariffs. The peak cooling load can be shifted from peak periods to off-peak periods by cold heat storage. Peak cooling loads of the buildings usually coincide with a very short time during the day. High-capacity cooling systems are used to meet this short-term but needed load. As the capacities of the cooling systems increase, their costs increase significantly. By decreasing the peak load with cold energy storage systems, the cooling system capacity required for space can be reduced. Thus, the size, capacity, and cost of the cooling systems are reduced. As mentioned earlier, cold heat storage systems have important economic advantages. Due to these advantages, it has an increasing interest and application area.

The initial investment costs and operating and maintenance costs of the cooling systems required for space can also be reduced with well-designed cold thermal energy storage systems. If one or more of the conditions listed in the following apply to a location, it may be considered attractive to use the cold thermal energy storage system for that location. If

- electricity tariff changes drastically during the day,
- electricity unit prices are expensive,
- the average cooling load is significantly less than the peak cooling load,
- installing cold storage systems is encouraged,
- increased capacity in AC is required,
- a new building is being built, or
- newer cooling system to be installed in place of the old cooling system.

2.9 Ice Storage Systems

In cold heat storage systems, chilled water, water—ice, and eutectic salts are widely used as storage materials. Water/ice is widely used as a storage medium in the cold heat storage systems because of its high latent heat capacity in phase change. In Fig. 2.22, the latent and sensible heat changes of water are shown. Water/ice phase change requires approximately 16 times higher heat change than 5°C of temperature change. Also, other significant features of water in heat storage material are easy availability, nontoxicity, cheap, and easy packaging. When water is used as a storage medium in a cold heat storage application, the system is called the ice storage system or ice thermal energy storage system.

The cooling load of the buildings is stored for later use by taking advantage of the latent heat required or released during the water/ice phase change in the ice storage systems. The main purpose of the ice storage systems is to reduce the energy consumption costs of AC systems by benefiting electricity tariffs. The working principles of ice storage systems are similar to other energy storage systems: energy

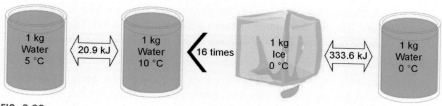

FIG. 2.22

The latent heat change of 1 kg of ice/water phase change, and the sensible heat change of 1 kg water for 5°C temperature difference.

charging, storing, and discharging periods. In the charging period, ice is produced during the off-peak tariff hours which the electricity unit price is the cheapest during the day. When the off-peak period is completed, the produced ice is kept until it is used in the insulated storage tank in the storing period. During the discharging period, the cooling demand of the building is carried out by using the produced ice during the midpeak and on-peak hours. Thus, the AC of the buildings is done at a lower cost. Besides lowering the operating costs of ACs, the main advantages of ice storage systems can be listed as follows:

- Reduce the operating costs of ACs: By reducing the use of AC systems during the hours when the electricity unit price is high, it reduces the cooling costs of the buildings. The economic performance of an ice storage system depends on the price change and distribution of the electricity tariff during the day.
- Shift the peak cooling loads to off-peak hours: Peak cooling loads correspond to several hours of the day. Thanks to ice storage systems, it will be possible to shift peak cooling loads to off-peak hours. Thus, it also helps to reduce the peak loads by providing a more balanced consumption of energy.
- Reduce chiller capacities: AC systems are designed according to the peak cooling loads of the buildings. Thanks to the ice storage systems, the capacity required to meet the peak load can be reduced by shifting these peak loads to off-peak hours.
- Provide chillers to work with higher the coefficient of performance (COP): AC systems are used extensively in the summer months. Especially due to the increasing outdoor temperature, solar radiation, and relative humidity in the afternoon, the working environment to which the chillers are exposed significantly reduces the COP of the chillers. Thanks to the ice storage systems, by using the chillers at night, the COP values of chillers can increase significantly.
- Reduce the capacities and sizes of auxiliary equipment such as a transformer, cable, generator, and so on required for building power: Lowering the peak cooling load provides a lower power demand from the grid. Thus, power can be provided to the building with a lower cable cross section. It will also be possible to reduce the generator capacity.
- Meet emergency demands: Cooling is a significant issue in buildings such as medical laboratories, data-processing centers, supercomputer facilities, and so

on. In such places, generator groups are used to continue cooling in cases of power failure. There are backups of these generator groups too. Ice storage systems can be used to reduce the number of generator groups or provide more effective cooling backup.

In addition to the benefits listed before, it helps to consume energy more efficiently because it provides more efficient operation of chillers and helps to reduce the energy peak load. Due to the higher energy efficiency provided, ice storage systems contribute to reducing fossil fuel consumption and carbon emissions; thereby they help to eliminate global warming effects. In addition, reducing the peak electricity consumption will also decrease the amount of imported energy, thus contributing to the reduction of the country's current account deficit due to energy imports.

The ice storage utilization strategy, also called utilization scenario, is defined as how much energy is stored and how stored energy is used. Several utilization strategies are available for ice heat storage systems. When only the storage capacity is considered, the strategies are classified as full storage and partial storage. However, it is significant how the stored energy will be used for the economic performance of the system. The ice storage utilization strategy is directly related to the initial investment cost of the system, saving amount, and payback period. Therefore, ice storage utilization strategy should be determined and applied to increase benefits. In Section 2.4.2 and Fig. 2.7, common ice storage utilization strategies are introduced. Also, the effect of ice storage utilization strategies is discussed in case studies.

While ice is produced and stored in the storage tank in the static system, in dynamic system, ice is produced outside the storage tank and then sent to the tank. Although dynamic ice storage systems have higher energy storage density and thermal performance than static systems, they have a more complicated system structure. Therefore, static ice storage systems are widely preferred in practical applications. Ice storage systems are classified as follows:

1. Static ice storage systems
   a. Encapsulated ice storage systems
   b. Ice-in-tube ice storage systems
   c. Ice-on-coil ice storage systems
2. Dynamic ice storage systems
   a. Ice slurry ice storage systems
   b. Ice harvester ice storage systems

2.9.1 Encapsulated Ice Storage Systems

In encapsulated ice storage systems, there are many ice capsules in the ice storage tank. The volume of capsules is very small compared with the volume of storage tank. The schematic view of the encapsulated ice storage systems is shown in Fig. 2.23. A typical encapsulated ice storage system consists of a storage tank, chiller, ice capsules, heat exchanger, heat transfer fluid, and pump. Generally, there

Chiller

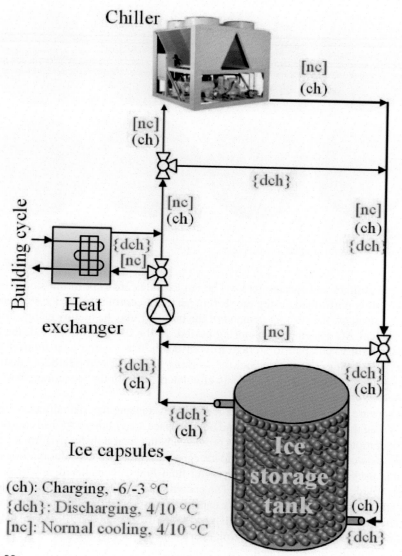

Building cycle

Heat exchanger

Ice capsules

(ch): Charging, -6/-3 °C
{dch}: Discharging, 4/10 °C
[nc]: Normal cooling, 4/10 °C

FIG. 2.23

The schematic view of an encapsulated ice storage system.

are thousands of ice capsules in the storage tank, and a heat transfer fluid, typically glycol ethylene solution, flows throughout the ice capsules. The storage capacity is determined by the number of capsules. In the charging period, the heat transfer fluid that is cooled to a temperature below 0°C is pumped to ice storage tank. Thus, water inside the ice capsules starts to solidify. Heat transfer fluid is circulated

FIG. 2.24

The views of several studied ice capsules.

Modified from Ref. [11].

between chillers and storage tanks. The ice capsules are kept in the storage tank until cooling performed during the storing period. According to the ice storage utilization strategy, the cooling demand of the building may be met by using chillers, as a typical AC system in the storing period. In the discharging period, the heat transfer fluid is pumped to storage to cool down. The cooled heat transfer fluid goes to the heat exchanger to cool the working fluid used in the building distribution system. The heat transfer fluid is circulated between the ice storage tank and the heat exchanger.

There are many different forms of ice capsules: rectangular, cylindrical, or spherical. When the ratio of the volume and outer surface area of these forms is taken into consideration, the spherical capsules provide a better heat transfer performance that occurs between heat transfer fluid and ice capsules. Spherical capsules provide more flexible packaging. When the heat transfer fluid flow is considered, spherical capsules provide a lower pressure drop. The encapsulated ice storage tank is considered, as a porous medium as the volume of capsules is too lower than the storage tank. In the encapsulated ice storage systems, the geometrically modified ice capsules are commonly used to increase the thermal performance. In Fig. 2.24, several commercial ice capsules are illustrated.

Ice capsule is also called ice ball or ice sphere. Ice capsules are filled with pure water. An air gap is left in the ice capsules to tolerate the volume changes during the water–ice phase change. In addition, some additives can be added into the water to reduce the effect of solidification (supercooling or subcooling) under freezing temperature. These additives prevent water from falling below 0°C without solidification during freezing. The effect of solidification under the freezing temperature increases the amount of energy required for icing and the solidification time. Therefore, it is desired to avoid this negative effect by adding additives.

2.9.2 Ice-On-Coil and Ice-In-Coil Ice Storage Systems

In this static ice storage system, the solidification takes place on the tube bank or inside the tube, depending on the type of the system. Fig. 2.25 illustrates a typical ice-on-coil ice storage system. In this system, there are tube banks in the storage tank and the tank is filled with water. Heat transfer fluid flows through the pipes. In the charging period, heat transfer fluid flows through the tube bank at a temperature below 0°C. Thus, the water inside the tank starts to solidify. The charging period is run during electricity off-peak tariff hours. In the discharging period, heat transfer fluid flows through the tube banks at a temperature higher than the melting temperature of the ice. Heat transfer fluid is cooled by transferring its energy to the ice in the tank. The cooled heat transfer fluid is pumped to the heat exchanger to cool the working fluid circulated in the building distribution cycle.

In ice-on-coil systems, the arrangement of the tube banks has the biggest impact on system performance. The distance between the tubes is important for the control of icing. According to the amount of energy to be stored, the diameter of the pipes, the distance between them, their arrangement (ordered or shifted), and the length of the pipes should be well designed. The ice-on-coil systems, also called the ice bank systems, can be used in low- and high-capacity systems.

The ice-in-coil system has the same system structure as the ice-on-coil. The only difference is the place of the ice. In the ice-in-coil system, ice is produced in the tubes and heat transfer fluid flows over the tube bank. Ice-in-coil systems are not widely used in practical application as it is hard to control the solidification and melting processes and requires high-strength tubes. They are generally preferred for low-capacity ice storage applications.

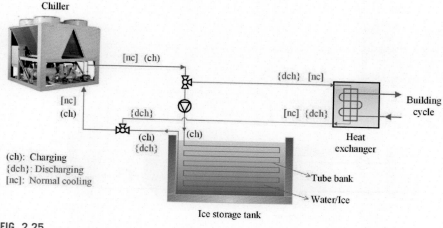

FIG. 2.25

The schematic view of an ice-on-coil ice storage system.

2.9.3 Ice Slurry Ice Storage Systems

The ice slurry system is a versatile cooling medium and ice storage system. The biggest advantage of the ice slurry is that it is pumpable. In addition to the cooling capacity of the ice slurry, the transport properties (pumpable) can be adjusted according to additives. Basically, an ice slurry consists of water, glycol ethylene, and mineral salts. At 20%–25% ice concentration, ice slurry flows like traditional chilled water and provides five times the cooling capacity than water. At an ice concentration of 40%–50%, an ice slurry stream shows dense muddy properties. In 65% –75% concentration, ice slurry is in ice cream consistency. When ice is produced in dry form (100% ice), it takes the form of nonstick pouring ice crystals that can be used directly in various products and processes (TES Book). The schematic view of an ice slurry system is shown in Fig. 2.26.

Ice slurry is a crystallized water–based ice solution that can be pumped and provides a secondary cooling medium for cold thermal energy storage, while enough liquid remains to pump. It provides five to six times more cooling capacity while flowing like traditional chilled water. That is, the ice mud system is a dynamic ice storage system type that offers a pumpable characteristic advantage over other types of dynamic systems. In the ice slurry ice storage system, the solidification takes place outside the ice storage tank like in every dynamic ice storage system. The ice produced in the system is sent to the ice tank via the pump. In the discharging period, the ice slurry in the tank provides the cooling of the working fluid used in the building distribution systems in a heat exchanger. Ice slurry systems have the following advantages:

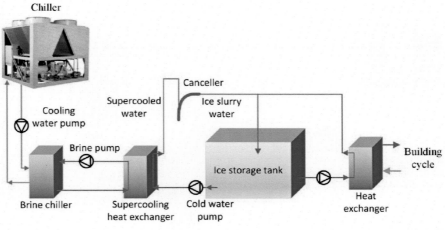

FIG. 2.26

The schematic view of the ice slurry system.

- High energy efficiency
- Cost-effective tank design
- Flexible tank design
- Ability to respond to higher capacities in a short time

Despite the positive properties listed before, ice slurry ice storage systems are not as common as static systems because the initial investment costs are quite high compared with static systems. The system has a more complex structure than static ice storage systems. Therefore, it is generally preferred in places with high storage capacity and a higher energy demand rate.

2.9.4 Ice Harvester Ice Storage Systems

In an ice harvester ice storage system, ice is obtained on plates integrated into the evaporator of the chiller. The schematic view of an ice harvester ice storage system is demonstrated in Fig. 2.27. In an ice harvester ice storage system, ice is obtained on plates integrated into the evaporator of the chiller. The water to be solidified is circulated periodically between the storage tank and the surface of the evaporator. Ice is obtained on the plates in the evaporator during the circulation of the water. The hot gas is sent to the inner surface of the evaporator plates to harvest the ice obtained on the surface. The ice produced is transferred to the ice storage tank located under the evaporator.

Here, water, and hot gas run cyclically. The biggest advantage of the ice harvester ice storage system do not contain a heat transfer fluid, and ice packaging (capsules or tubes). Ice-on-coil systems can also be used as an ice harvester system.

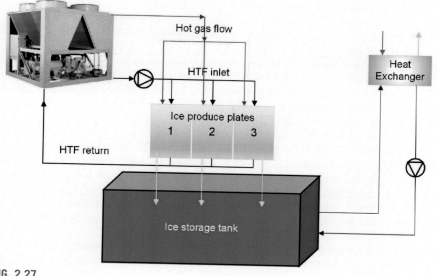

FIG. 2.27

The schematic view of the ice harvester ice storage system.

However, a large surface area is needed for ice produce and hot gas contact. Therefore, plates are preferred for ice harvester systems. The discharging period is similar to other ice storage methods. The stored ice is used for building cooling demand.

2.10 Closing Remarks

In this chapter, the fundamentals of sensible heat and latent heat storage systems are discussed. The advantages and disadvantages of heat storage techniques are presented with examples from practical applications. Common storage mediums used in sensible and latent heat storage systems are listed. The advantages and disadvantages of PCMs are also presented to readers. Cold heat storage systems and their applications are given in detail with practical applications.

References

[1] T.L. Bergman, A.S. Lavine, F.P. Incropera, D.P. Dewitt, Fundamentals of Heat and Mass Transfer, John Wiley & Sons, Ltd, 2011.

[2] D. Erdemir, N. Altuntop, Effect of thermal stratification on energy and exergy in vertical mantled heat exchanger, Int. J. Exergy 20 (2016), https://doi.org/10.1504/IJEX.2016.076681.

[3] I. Dinçer, M.A. Rosen, Thermal Energy Storage: Systems and Applications, second ed., 2010, https://doi.org/10.1002/9780470970751.

[4] M.Y. Haller, C.A. Cruickshank, W. Streicher, S.J. Harrison, E. Andersen, S. Furbo, Methods to determine stratification efficiency of thermal energy storage processes — review and theoretical comparison, Sol. Energy 83 (2009) 1847–1860, https://doi.org/10.1016/j.solener.2009.06.019.

[5] Borehole Thermal Energy Storage, n.d. https://www.dlsc.ca/borehole.htm. (Accessed 15 July 2020).

[6] I. Dincer, M.A. Ezan, Heat Storage: A Unique Solution for Energy Systems, Springer, 2018, https://doi.org/10.1007/978-3-319-91893-8.

[7] E. Özrahat, S. Ünalan, Thermal performance of a concrete column as a sensible thermal energy storage medium and a heater, Renew. Energy 111 (2017) 561–579, https://doi.org/10.1016/j.renene.2017.04.046.

[8] S.D. Sharma, K. Sagara, Latent heat storage materials and systems: a review, Int. J. Green Energy 2 (2005) 1–56, https://doi.org/10.1081/GE-200051299.

[9] D. Erdemir, A. Ozbekler, N. Altuntop, Experimental investigation of the effects of phase change material usage in A vertical mantled hot water tank, Solar Energy (2021) (Submitted paper, not accepted by journal).

[10] D. Erdemir, N. Altuntop, Experimental investigation of phase change material utilisation inside the horizontal mantled hot water tank, Int. J. Exergy 31 (2020) 1–13, https://doi.org/10.1504/IJEX.2020.104722.

[11] D. Erdemir, Numerical investigation of thermal performance of geometrically modified spherical ice capsules during the discharging period, Int. J. Energy Res. 43 (2019) 4554–4568, https://doi.org/10.1002/er.4585.

Energy Management in Buildings

3

3.1 Introduction

Buildings are recognized as one of the critical sectors as they are responsible for 36% of the energy consumed in the world. Such a high energy demand in return results in some high amounts of greenhouse emissions and appears also to be responsible for about 39% of carbon emission [1]. Since they take place in everybody's life, everyone is responsible for this energy consumption sharing. This also brings a big advantage since individuals can contribute energy savings, more efficient energy usage, and effective energy management. When the number of buildings in the world, which is more than 6.2 billion, and the world population, which is more than 7.8 billion is taken into consideration, a small contribution provided by individuals may turn into a giant effect in energy savings, more efficient energy use, and effective energy management. Also, individual actions in energy-efficient applications require lower investments and service costs compared with high-capacity applications. When the energy-efficient systems and actions would like to be supported by governments or organizations, it is important to provide the equality of opportunity and equal distribution. Therefore, buildings and people who live in buildings are a significant part of energy-efficient systems and actions.

Buildings are basically classified into two categories as a common approach, namely residential and commercial. A residential building is defined as the building that provides more than half of its floor area for dwelling purposes. In other words, the residential building provides sleeping accommodation with or without cooking or dining or both facilities. Residential buildings, although there are many more different varieties, may be categorized as follows (e.g. Ref. [2]):

- individual houses or private dwellings,
- lodging or rooming houses,
- dormitories,
- apartments, and
- hotels.

Commercial buildings can be defined as buildings that are used for commercial purposes. All buildings except for residential purposes such as supermarkets, office buildings, shopping centers, banks, hospitals, and so on are classified as commercial buildings.

Heat Storage Systems for Buildings. https://doi.org/10.1016/B978-0-12-823572-0.00004-7

Residential and commercial buildings are places people spent a long time. Therefore, buildings play a key role to raise people's living standards and comfort levels. Natural gas, electricity, coal, and fuel oil are common fossil fuels used in the buildings. As a renewable source, solar heating and PV systems are commonly applied to the buildings. The low-capacity wind turbines are also used in buildings or communities as a renewable energy source. Geothermal is used for heating and power generation purposes.

When the amount of energy consumed and the energy demand profile in the buildings is considered, energy storage systems have great potential to manage energy consumption. Energy storage systems can reduce energy consumption costs, balance energy demand profiles, and reduce peak energy demands. Heating and cooling are responsible for almost half of the energy consumed in buildings. Therefore, heat storage systems are a unique option for the buildings as heat is the last type of heat to be used. Heating and cooling demands of the buildings can be stored for later use to reduce energy costs and device capacities.

In this chapter, to determine the potential of heat storage systems, first, the energy consumption profile in residential and commercial buildings is investigated. The energy sources used in buildings are discussed. Then, the energy systems used in the building are summarized to readers.

3.2 Energy Demand Profiles in Buildings

Driven by the ever-expanding economy, ever-rising population, living standards, and comfort levels at the global scale have remarkably increased the energy consumption in the world, as stated in Fig. 3.1. Like in any sector, the energy consumed in the residential and commercial buildings is ever increased. Future projections, as predicted by many, show that energy consumption will continue to increase in all sectors in the future. Fig. 3.1 illustrates that the share of energy consumption in residential, commercial, and public service sectors is 32% of total energy consumption in 1990. This ratio has increased up to 34% until 2001. In 2017, it has reduced to 29%. The decrease between 2001 and 2017 has occurred due to the wide use of energy-efficient devices in the buildings. To increase the use of energy-efficient devices and guide the customer in the market, many applications have been carried out, such as energy-efficient labels. In Fig. 3.2, the energy labels used in refrigerators, washing machines, and vehicles are shown. Such practices are both informative and encouraging to people. Since devices used in daily life such as TV, computer, refrigerator, air conditioner, washing machine, heater, and so on can be easily selected according to their energy-efficient situation, they have wide usage. Thus, it is possible to reduce energy consumption in residential buildings. Extending these labeling and grading applications to devices and applications, which consume energy, will play an important role in reducing energy consumption.

In the United States, Energy Star symbol, which is the government-backed symbol for energy efficiency, is used for providing simple, credible, and unbiased

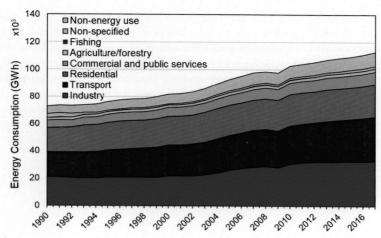

FIG. 3.1

Change of energy consumption for all sectors in the world.

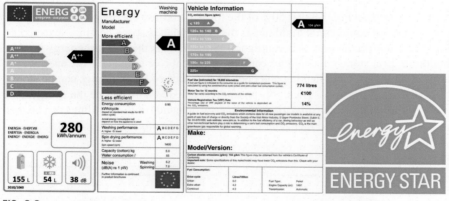

FIG. 3.2

Energy labels used in the European Union.

Adapted from Refs. [3,4].

information that consumers and businesses rely on to make well-informed decisions. Energy Star, and its partners that are thousands of industrial, commercial, utility, state, and local organizations, have provided saving in residential and commercial sectors more than 4 trillion kilowatt-hours of electricity and achieve over 3.5 billion metric tons of greenhouse gas reductions, equivalent to the annual emissions of more than 750 million cars. In only 2018, Energy Star and its members have helped to save $35 billion in energy costs in the United States [4].

In addition to energy labels, the exergy labels must be designed and applied to energy systems in the near future. Exergy labels will provide to choose a system more accurately. It is possible to select the optimum energy source for a system. Exergy labels can define which energy source is suitable for the energy system.

Fig. 3.3 illustrates the disaggregated end-use electricity demands for residential and commercial buildings. The energy demand items for both residential and commercial buildings are almost the same since both provide a place for people to live. And people's needs are almost the same in both buildings. Energy consumed in buildings is used for the following purposes:

- space heating,
- space cooling,
- lighting,
- water heating,
- ventilation,
- refrigeration,
- electronics,
- cooking, and
- other purposes.

It is clear from Fig. 3.3 that 50% of the form of end-use energy consumed in both residential and commercial buildings is the heat which is space heating, space cooling, hot water production, and ventilation. Since heat is the end-use energy type in buildings, heat storage techniques can play a critical role to manage the energy systems in buildings. Both heating and cooling demands of the buildings can be stored by heat storage methods. Thus, it is possible to considerably reduce the energy cost and system's equipment capacities. The energy demands of the buildings can be met in a sustainable way and managed in optimum conditions.

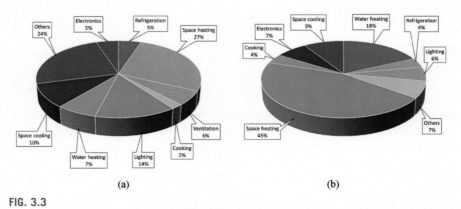

(a)   (b)

FIG. 3.3

End-use energy consumptions: (a) residential and (b) commercial buildings.

Data from Ref. [5].

Fig. 3.4 demonstrates the energy use shares by various sectors in Canada. In Fig. 3.4a, it is clearly seen from the change of energy use in Canada from 1990 to 2017 that residential and commercial—institutional buildings are responsible for one-third of gross energy use. While the energy use in the residential sector has increased from 1425 to 1508 PJ (5%), it has increased from 746 to 1030 PJ (38%) in the commercial and institutional sectors. In Fig. 3.4b, the share of sectoral energy use is shown for 2017. In 2017, 40% of total energy use is in industrial sectors.

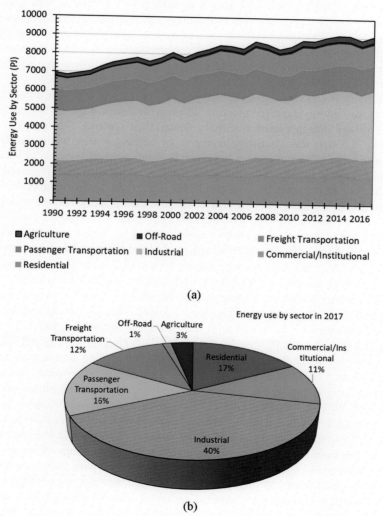

FIG. 3.4

Energy use by sectors in Canada: (a) change from 1990 to 2017 and (b) share in 2017.

Data from Ref. [6].

Residential and commercial/industrial sectors are responsible for 17% and 11% of total energy use, respectively. For transportation, 28% of total energy consumed has been used.

The share of energy demand items seen in Figs. 3.3 and 3.4 is for the average of residential and commercial buildings in the United States and Canada. These rates can be different for specific buildings such as hospitals, airports, shopping centers, supercomputer centers, and so on. In the buildings where heating and cooling are critical, heat storage systems can play a significant role to reduce peak loads and equipment sizes. When the number of buildings and their energy consumption rates are considered, heat storage systems are a unique solution providing to meet energy demand in an optimum way. They can provide applications spread to the base. This also provides equal opportunities and sharing for individuals both the investments and the benefits obtained.

Electricity is one of the most used energy types in buildings as it is easily converted to any energy type for meeting different demands. Fig. 3.5 illustrates the hourly average disaggregated electricity demand for Ontario, Canada. It is clear from Fig. 3.5 that the electricity demands of buildings peak when the devices consume energy used at the same time. Cooling and heating demands of buildings are one of the highest energy consumption items in residential and commercial sectors. The cooling demands increase due to the increase in outdoor temperatures, the effect of global warming, prolonged time spent in closed areas, and the widespread use of devices and equipment that increase heat gain. Also, the air-conditioning (AC) period is extended. Due to both reasons, cooling is a serious cost for buildings. Cooling loads have great impact on electricity demand compared with heating loads.

This because, while many energy sources such as coal, natural gas, fuel oil, electricity, and so on can be used for heating applications, electricity is used in AC systems as an energy source. As heating and cooling systems consume large amounts of energy and have a large impact on the electrical peak load, it is critical and important to reduce the electrical peak load with heat storage systems (e.g., Ref. [8]).

Fig. 3.6 demonstrates the disaggregated electricity consumption by end use in the United States in 2018. As seen in Fig. 3.6, heating and cooling demands are responsible for about 57.77% of total electricity consumption. In other words, space heating, space cooling, hot water, furnace fans, and boiler circulation pumps are more than almost 60% of total electricity consumption. In other words, space heating, space cooling, hot water, furnace fans, and boiler circulation pumps are more than almost 60% of total electricity consumption. It is clear that heating and cooling purposes are the main reason for the energy consumption in buildings. Also, as seen in Fig. 3.5, heating and cooling demands have fluctuating behavior throughout the day.

As seen from figures as presented in this section, heating and cooling systems and the equipment used in these systems are responsible for almost half of the total energy consumption in buildings in residential and commercial sectors. These demands bring a serious cost with them. Additionally, the trend of change in heat and cooling demands shows fluctuating behavior throughout the day. Since the heating and cooling systems are generally used at the same time, they have a great

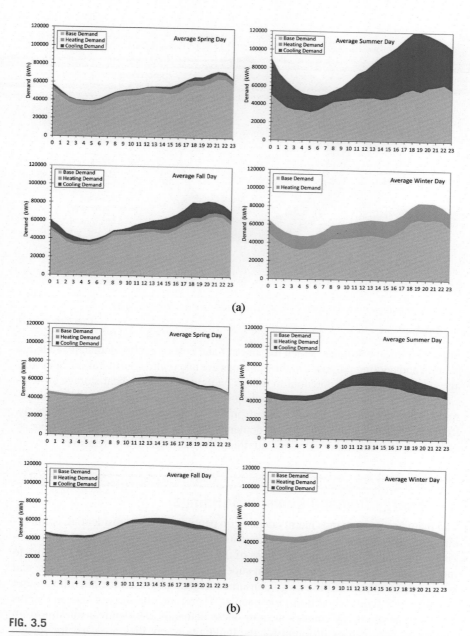

FIG. 3.5

Average disaggregated electricity demands for (a) residential buildings and (b) commercial buildings.

Data from Ref. [7].

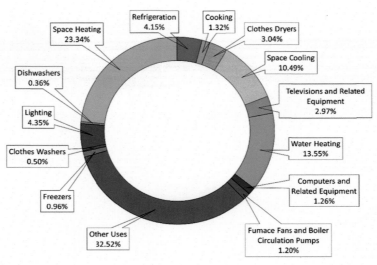

FIG. 3.6

Residential energy consumption by end use in the United States, 2018.

Data from Ref. [9].

impact on grid peak loads. When the amount and cost of energy demands related to heating and cooling purposes in the buildings are taken into consideration, heat storage systems have a big potential to reduce energy costs. Also, heat storage systems can play a critical role in shifting cooling loads from peak periods to off-peak periods. The energy supply and demand profiles can be managed more efficiently. Reducing heating and cooling peak demands help to reduce the size and capacity of heating and cooling devices and auxiliary devices used in the systems.

3.3 Energy Demand Items in Buildings

The main energy demands mainly depend on people's needs in residential, commercial, and institutional sectors as buildings are places where peoples live. In commercial sectors, although the main propose of the use of buildings is to produce work to meet people's needs such as banking, supermarket, shopping centers, airports, and so on, the most part of the energy consumed is generally used for providing better working and living conditions for people. Therefore, the energy demand items and the type of energy sources meeting these demands are identical for almost all buildings. They can be different for the buildings where they are used for special purposes such as supercomputer centers, stem cell, or embryo centers as they need special needs such as cooling, ventilating, and so on. The energy demand items in buildings can be listed as follows:

- space heating,
- space cooling,
- ventilation and fresh air,
- hot water,
- lighting,
- cooking, and
- appliances and equipment.

As mentioned earlier, space heating and cooling are responsible for almost half of total energy consumption of the buildings as they are a significant need for people's comfort conditions. Cooling energy use in buildings has doubled since 2000, making it the fastest growing end use in buildings, led by a combination of warmer temperatures and increased activity.

Many energy demand items have been joining our lives with developing technologies and devices invented. A few years ago, electric vehicles and their energy demands were not an essential issue for people. However, today, electric vehicles and their charging demands are one of the critical issues. As another example, with the COVID-19 pandemic, fresh air has become one of the most significant needs for people's health and safety for indoors. It brings a massive amount of energy consumption with it, too. However, the buildings' infrastructures do not meet additional energy demand items and increases in the existing systems' capacities.

Heat storage systems are a convenient method to make room for additional energy demand items and capacity increases. Heating and cooling loads of the building can be shifted to off-peak periods. Thus, energy demand profiles are managed efficiently by shaving peak energy demand profiles.

3.4 Energy Sources Used in Buildings

Although many energy sources are used for meeting the energy demands of all sectors, from coal to nuclear and from solar to wind, the types of energy sources used in buildings are limited due to physical conditions, safety concerns, and so on. When the amount of consumed energy in buildings, which are responsible for almost 40% of the total consumption, is taken into consideration, the limited number of energy sources causes overload in specific energy sources. The typical energy sources and fuels used in buildings can be listed as follows:

- electricity,
- coal,
- natural gas,
- fuel oil,
- biomass,
- gasoline, and
- diesel.

In addition to the traditional energy sources and fuels, three renewable energy sources are directly used, as the most common resources, for meeting the energy needs in buildings:

- solar,
- wind, and
- geothermal.

When the buildings' needs stated previously in Figs. 3.3 and 3.6 are considered, while electricity is the most widely used energy source, natural gas is the most widely used fuel in the buildings. Electricity is the only energy source for electric devices (television, computer, oven, microwave, dishwasher, dryer, etc.) and HVAC (heating, ventilation, and air-conditioning) systems. Natural gas is used for heating and cooking purposes. Coal and fuel oil are used for heating. Gasoline and diesel are used for fuel power generators used in power outages and remote areas.

3.5 Building Energy Systems

Although many devices and systems are used in the buildings, the type of energy sources and fuels are limited due to the physical conditions, safety concerns, and storage issues. Electricity devices are used in people's daily life, such as dishwashers, washers, dryers, TVs, computers, ovens, microwaves, kettles, and so on. Although these devices are widely used in daily life, they are not responsible for massive energy consumption. Similarly, elevators, lightings, and security systems are also systems that do not cause high energy consumption. When the amount of energy consumption rates are considered, the energy systems in the buildings can be classified as follows:

- HVAC systems,
- heating systems,
- AC systems,
- heat pumps,
- pumps and hydrophores,
- hot water systems,
- power generators (diesel, gasoline, battery), and
- power generators for electricity generation (gas turbines).

Although it is not classified here, many subsystems, devices, and equipment are used in these systems such as fans, pumps, motor-driven valves, heat exchangers, and so on. Each system, subsystem, and equipment will be introduced in the following section, chapters, and case studies. In this section, HVAC systems are briefly introduced to readers.

3.6 Heating, Ventilation, and Air-Conditioning Systems

HVAC systems are used for artificially creating and maintaining an indoor environment. Artificial environment conditions are provided by operating one or a few of heating, AC, humidifying, dehumidifying, fresh air systems according to existing indoor and outdoor environment conditions. The main purpose of an HVAC system is to maintain environment conditions in a closed space that are favorable to

* provide people's comfort and health (called thermal comfort conditions), and/or
* maintain certain environmental conditions for the requirements of special products, processes, or industrial processes.

The illustration of thermal comfort conditions on the psychrometric chart for people is given in Fig. 3.7. It is aimed to keep thermal comfort conditions that are shown in the blue zone in Fig. 3.7. The processes demonstrated with arrows in Fig. 3.7 should be performed to provide and maintain the desired thermal comfort conditions.

Also, various environmental and thermal conditions are required for specific products, processes, or industrial processes. For instance, stem cell and embryology processes require very certain environmental conditions. Also, a relatively dry environmental conditions should be ensured for libraries and museums. Also, in the production of various products such as electronic chip, fabric, tempered glass, and so on, a special environmental condition, that is, cold and dry, should be provided. Symphony orchestra buildings should be kept at the same environmental conditions

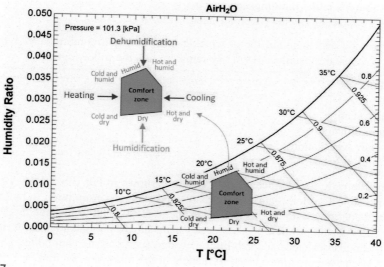

FIG. 3.7

Thermal comfort zone for peoples as illustrated in the psychrometric chart.

not to go out of the chords of music instruments. Lastly, environmental conditions of greenhouses should be adjusted according to the plant to be grown.

Both environmental and thermal comfort conditions items can be listed as follows:

- temperature,
- humidity,
- air motion (speed),
- air cleanliness and quality, and
- noise (sound, electricity, and magnetic).

The atmospheric conditions and physical characteristic of the building affect the environmental and thermal conditions in a space. Atmospheric conditions, also called outdoor conditions, can be listed as follows:

- temperature,
- relative humidity,
- wind speed and direction,
- precipitation, and
- insolation.

The physical characteristics of building affecting the environmental and thermal conditions can be itemized as follows:

- the number of floors,
- buildings materials,
- windows,
- heat sources in the buildings, and
- ventilation requirements (fresh air need).

To provide and maintain the acceptable environmental conditions in a space, HVAC systems are installed. HVAC systems provide the following:

- heating,
- cooling,
- humidification,
- dehumidification,
- fresh air (ventilation), and
- air filtering.

The schematic for a typical HVAC system is shown in Fig. 3.8. HVAC systems are responsible for more than two-thirds of total energy consumption in buildings. To reduce energy consumptions and costs due to HVAC and to manage the energy supply and demand profiles, heat storage systems have a big potential. Heat storage systems can also enhance adaptability renewable energy systems to building applications. Each system is introduced to readers in the following sections.

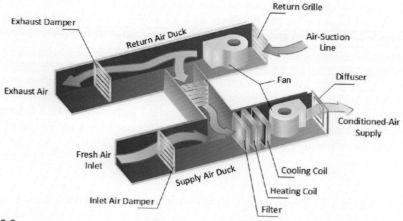

FIG. 3.8

The schematic view of a typical HVAC system. *HVAC*, heating, ventilation, and air-conditioning.

3.6.1 Heating Systems

Building heating systems are used for increasing the temperature of a space in buildings or industrial processes. A heating process is primarily achieved by

- free convection (electric heaters, radiator, etc.),
- forced convection (fan coils, forced air over a heat exchanger in furnace, etc.), and
- radiation heat transfer (radiators, electric radiative heaters, etc.).

In heating systems, heat is generated by converting an energy source into heat or using a fuel. Heat is used for increasing temperature of air to be blown to the space or a heat transfer fluid to be pumped to a heat exchanger placed in the space.

When the building needs are considered, heating systems are used for controlling the indoor temperature in cold weather conditions. According to the weather conditions, development level of the countries, and fuel types that countries have, the heat generation method and fuels vary. Today, although natural gas and electricity are used as source for heating purposes in the developed countries, coal, wood, and other biomass-based fuels are still used in undeveloped and developing countries. In design of the heating systems, the peak heating load of the building is determined by calculating heat loss from the building. Types of heat losses from a building can be classified as follows:

- heat conduction,
- heat convention, and
- air leakage.

Note that both heat conduction and convection define the heat losses that occur from physical boundary of the buildings such as walls, windows, and so on. Heat loss with heat conduction and convection can be calculated as follows:

$$\dot{Q}_{loss} = A \, K \, \Delta T$$

Here, $\dot{Q}_{loss}$ (kW) denotes the heat transfer rate that occurs from indoor to outdoor. K (kW/m°C) is the overall heat transfer coefficient of the building materials. A is the surface area of the building materials, and ΔT is the temperature difference between indoor and outdoor. The overall heat transfer coefficient (K) can be calculated with the following equation:

$$\frac{1}{K} = \frac{1}{h_{in}} + \frac{x_1}{k_1} + \ldots + \frac{x_n}{k_n} + \frac{1}{h_{out}}$$

where h_{in} and h_{out} denote the heat conduction coefficients between the surface of the building material and surrounding air. $x_1, x_2, \ldots, x_n$ are the thicknesses of the materials, and $k_1, k_2, \ldots, k_n$ are the heat conduction coefficients of the materials. Mechanism of the heat transfer in building materials is shown in Fig. 3.9.

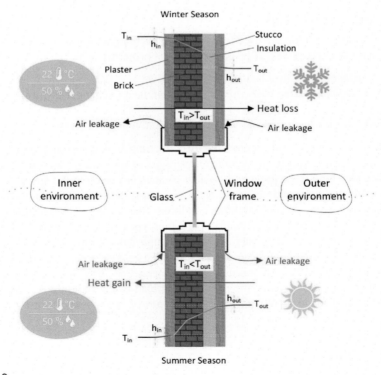

FIG. 3.9

Heat transfer mechanisms in the building materials.

The heat loss through air leakage is caused by the cold air leaking into the room from window and door openings due to the pressure difference between the outside air and the indoor air of the volume. It can be calculated as follows:

$$\dot{Q}_{leak} = (1/3.6)(a\,L)R\,H\,\Delta T\,Z_e$$

Here, "$\dot{Q}_{leak}$" is heat loss that occurs due to air leakage. "a" is the coefficient of leakage. "L" is the length of the door and windows. "R" is the wind permeability coefficient. "H" is the effective wind coefficient. "ΔT" is the temperature difference between indoor and outdoor surroundings. "Z_e" is the coefficient of the corner openings. In practical applications, the heat loss charts are used for systematically and accordingly calculating the total heat loss for a building. These charts are prepared and published by the Chamber or Society of Mechanical Engineers.

While calculating the total heat loss from a building, the data and information required are taken from the architectural project, meteorological statistics, and engineering standards. To determine the heat load of the building, the heat loss calculation for each room (volume) in the building should be performed. The capacity of the heater to be placed in the room is determined, based on the heat lost by the room. The heat loss of the building is calculated by adding up all heat losses calculated for each unit of the building. If individual heating is performed, a heater (floor heater, combi boiler, furnace) corresponding to the heat loss of the apartment is selected. If there is a central heating system for all buildings, boilers or cascade combi boiler systems are selected according to the total heat loss of the building. Heating systems can also be applied for all through campus or site. In this kind of systems, high-capacity boilers are used.

In the heating systems, air, water, or a heat transfer fluid such as ethylene glycol is heated in a furnace or boiler by burning a fuel such as natural gas, coal, propane, fuel oil, and so on. Electricity can also used for heating purposes. In electricity heating systems, the heat is directly converted into heat via resistances. In addition to conventional fuels and electricity, solar energy systems and geothermal sources can also be used for heating application. Solar heating systems are one the most common renewable energy applications. In solar heating systems, solar air or water collectors are used for converting the solar energy into the heat. In geothermal-based heating systems, the hot water in the reservoir is directly pumped to the buildings. A heat exchanger is used for transferring heat in the geothermal water into the building cycle. Today, heating systems can be assumed as mature systems. Each systems and possible heat storage applications are discussed in the following chapters and case studies.

3.6.2 Air-Conditioning Systems

Cooling is performed to reduce the temperature of the closed volumes, such as buildings, refrigerators, cold storage depots, and so on. When the cooling is done between ambient temperature and 1°C in buildings, it is called AC. The cooling can also be performed at various temperatures between 10°C and −40°C for cold storage, commercial, and industrial purposes. One of the biggest reasons for the need of cooling systems is to provide thermal comfort condition. Today, AC systems have become one of the most significant needs of people's daily life. Due to the rising living standards and comfort levels at the global scale and the increase in the time spent indoors, cooling of buildings has become a significant initial investment and operating cost item. The capacity and working duration of AC systems are growing due to increasing environment temperature, spending a more extended time indoors, and using many devices producing heat in buildings. The extended-use period and high capacity of ACs bring significant operating and initial investment expenses with them for buildings.

An AC system can be classified into two types: centralized systems and decentralized systems. While the system cools more than one zone in centralized systems, the system serves for one zone (a unitary system). According to the agent flow type, AC systems can be divided into three types: all-air, air–water, and all-water systems.

While determining the AC capacity for space or building, heat gains from the environment and produced in the space are calculated. Heat gains are calculated in a similar way in the heat loss. In heat gains, the outdoor temperature is higher than the indoor temperature. So, the direction of the heat transfer is reversed according to the heating case. In the cooling load calculation, the heat gains due to the device and system used in the space are taken into consideration. Also, heat gains due to the number of persons are added to the cooling loads. Lastly, cooling loads of the fresh air are added to the cooling load, and the total cooling loads of a unit in the building are determined by summing all cooling load items. If a decentralized system will be applied to the unit, a unitary AC system is installed according to the cooling load in the space considered. If the cooling will be performed for the entire building, a centralized AC system is installed in a proper place in or around the building. The heat transfer fluid cooled down in the AC system is pumped to the air handling unit in the spaces.

The main parts of a centralized AC system can be divided into two types: primary and secondary systems. The primary system converts an energy source to thermal energy for cooling and heating. The secondary system, also called the air handler, distributes heating, cooling, and fresh air (if required) to the building units. The following methods are used in AC systems:

- vapor compression mechanical cooling,
- absorption cooling,
- adsorption cooling,
- thermoelectric cooling system,
- magnetic cooling,
- paramagnetic cooling,
- cooling by vortex tube,
- humidified (evaporative) cooling,
- melt formation cooling,
- cooling by the expansion of gases,
- vacuum cooling,
- steam jet (ejector) cooling,
- sterling cooling system, and
- acoustic (sound) cooling,

Almost all AC systems used in building applications work with the principle of the vapor compression refrigeration cycle. Fig. 3.10 shows a typical single-stage

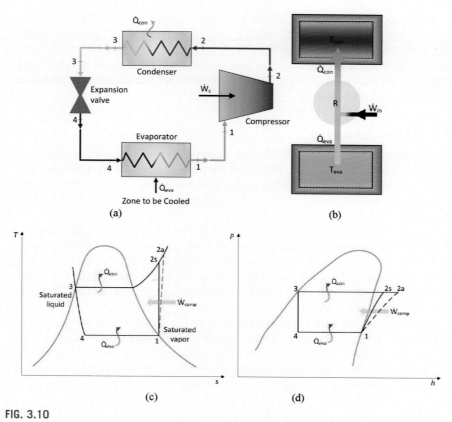

FIG. 3.10

(a) Schematic diagram, (b) illustration, (c) T-s and (d) P-h diagrams of a basic vapor compression refrigeration cycle.

vapor compression refrigeration cycle. A typical system consists of a compressor, condenser, expansion valve, and evaporator. The system components are connected by copper pipes to create a closed cycle. Here,

1-2 compression: where the work input is supplied externally to a compressor to compress the refrigerant. The process may be isentropic compression under ideal condition.

2-3 heat rejection: where refrigerant is cooled down for condensing. The temperature of refrigerant may reduce a desired temperature after being condensed. The process is isobaric under ideal condition.

3-4 expansion: where the liquid refrigerant is expanded to a desired pressure to become a saturated liquid and vapor mixture. The process is isenthalpic under ideal condition.

4-1 heat addition: where the liquid refrigerant evaporates by absorbing heat. The process is isobaric under ideal condition.

Each process in an ideal refrigeration cycle is shown in Fig. 3.10c and d in T-s and P-h diagrams, respectively. Ideal cycles are quite crucial for understanding and analyzing the systems and their working principles. However, under actual working conditions, though the cycle looks similar, the locations of the state points change. Examples for the ideal and actual refrigeration cycle are thermodynamically analyzed in the next chapter.

3.6.3 Heat Pumps

Heat pump systems are used for heating purposes though they work with the principle of the refrigeration cycle. Therefore, they are often called the reversed refrigeration cycle. Although heat pump cycles look like refrigeration cycles, the main difference is using them for heating. The schematic diagram of a basic heat pump is shown in Fig. 3.11.

Heat pumps are generally costlier to install than other heating systems such as furnaces working with coal, natural gas, fuel oil, and so on and electrical heater. However, when the long-term use period is considered, they can provide a huge amount of savings. They can also play a critical role in the electrification of heating purposes. When the electricity is met by renewables, the heating demand can be met in a more sustainable way.

In heat pumps, the most critical issue is to find an effective low-temperature energy source. The most common energy source for low- and medium-capacity heat pumps is atmospheric air. So, they are called air-to-air heat pumps. Water and soil are also used as low-temperature energy source. In recent studies, researchers and engineers are working on integrating the cold heat storage systems to heat pumps

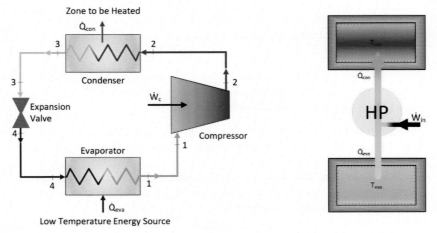

FIG. 3.11

Schematic diagram of a conventional heat pump system.

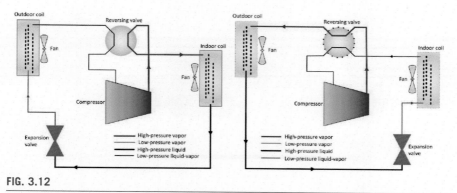

FIG. 3.12

The schematic diagram for a heat pump system used for heating and cooling purposes.

Modified from Ref. [10].

as the energy source. The coefficient of performance (COP) of heat pumps varies between 1.5 and 4, depending on the specific system used and the temperature of the low energy source.

Heat pumps and ACs have the same components as previously stated. It is possible to combine them by adding a reversing valve to the system. Thus, the cycle can be used for both cooling purposes in the summer and heating purposes in the winter. The reversing valve is used for directing the refrigerant to indoor or outdoor heat exchanger. A schematic for a heat pump used for both heating and cooling is shown in Fig. 3.12.

3.6.4 Cogeneration Systems

Cogeneration systems can be defined as energy systems that have the capability to produce two useful outputs simultaneously. They are unique techniques to benefit from an energy source in a more effective and sustainable way. As shown in Fig. 3.7, they are one the most significant branches of the smart energy portfolio introduced by Dincer [11]. In the past few decades, the use of cogeneration systems has increased due to their crucial benefits. Today, one of the most used types of cogeneration systems is the combined power and heat systems. They are also generally used for building applications as they meet both electricity and heating demands. In combined heat and power cogeneration systems, fuel or fuel blends are used in furnace or gas turbine, and then electricity is generated as the primary output. The waste heat is used for heating purposes such as space heating and domestic hot water. When the number of useful outputs is more than two, the system is called as the multigeneration system. Today, in the practical applications, it is possible reach six useful output effectively. However, when the building application is considered, cogeneration systems called the combined heat and power systems are the most common one. The electricity and heating demand of the building are met by these cogeneration systems.

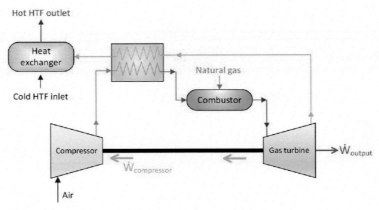

FIG. 3.13

The schematic diagram for the combined heat and power generation system.

Today, in the practical cogeneration applications in the buildings, gas turbines fueled with natural gas are used power generation, and the waste heat is used for heating purposes. Also, renewables can be integrated into the system as the primary or auxiliary energy source. The schematic diagram of a typical combined heat and power generation system is shown on Fig. 3.13. A typical Brayton cycle fueled with natural gas is used for power generation. Some of the electricity generated is used in the compressor. The waste heat is used for preheating of the air and heating applications. Today, many improvements have been applied to the cogeneration system to increase the system efficiency. Also, it is possible to integrate the renewables into the system.

Cogeneration systems are sometimes used for shaving peak loads. In some regions, a higher electricity tariff rates are applied to subscribers having a higher electricity consumption. To reduce electricity costs, integrating cogeneration system is one of the most suitable techniques. It is very important for hospitals, malls, and factories as they need both electricity and heat. Cogeneration system is also a critical solution when there is no or enough power infrastructure for a building. It is also an energy-efficient solution for campuses and sites where both power and heat are needed.

Heat storage methods play a significant role in managing the heat supply and demand profiles in the combined heat and power generation system. They are used for storing excessive electricity and heat produced by cogeneration system for later usage. They help to reduce capacity and size of them. Therefore, renewable energy and heat storage—integrated cogeneration systems can be cost- and energy-effective method to meet power and heat demand of the buildings. Individual systems can also help to manage the grid loads.

In the open literature, there are many studies related to cogeneration systems and their improvements. In the case studies, we have presented a few interesting applications of the cogeneration systems and their building applications.

3.6.5 Standby Generators

Power generators, also called standby generators, are used for meeting electricity demands in the case of power outages in all sectors. They start to generate electricity by sensing a power outage, automatically turning themself on. They can provide the power required within a few seconds. Thus, they prevent unsafe working conditions, loss of electronic data, and breakdown of the devices. Standby generators keep the essential systems up and running such as data centers, elevators, security systems, and so on. They also maintain the desired climatic condition as they are supplying the power needed by HVAC systems. Fig. 3.14 demonstrates the view of the diesel-fueled standby generator used in the Ontario Tech University. They get involved into power generation when there are power outages and provide the continuous power generations. Today, they are an essential part of the buildings due to certain needs on the buildings. They are also used for meeting power demands where there is no grid connection, such as communication towers and construction sites in rural areas.

In some countries, there are some attempts to reduce emissions due to fossil fuel−driven standby power generators. For instance, the California Public Utilities Commission is considering replacing the diesel generators with ammonia-driven ones for 2021 [12]. Here, it should be emphasized that ammonia technologies are a unique solution to transition to hydrogen system and economy.

In the past decade, battery systems are able to be used instead of standby generators. A battery system used in the communication tower is shown in Fig. 3.15. The capacity of the standby generator and battery systems is determined according to the energy demand of the building. A system with a capacity as much as the energy needed is established in the case of power outages. In some buildings serving for special services such as supercomputer, data centers, stem cells, embryology, and so on, there are backups of standby generators. These systems bring a high cost with them. In these kinds of buildings, the main need is to maintain the certain climatic condition in the buildings. Heat storage systems can play a critical role in

FIG. 3.14

The view of a standby generator group from Ontario Tech University, Canada.

FIG. 3.15

The view of the battery groups used instead of the generator.

reducing the capacity and cost of standby generators and battery systems when these systems are mainly used for HVAC systems. While heat storage systems can reduce the energy cost for normal days, they can also be used in the case of an emergency situation. Thus, the capacity of the standby generator and battery systems can be managed and reduced with the integration of heat storage systems.

3.7 Closing Remarks

Energy efficiency and management is a critical issue in the buildings as they are responsible for almost one-third of energy consumption in the world. Heat storage systems can play an important role in reducing energy costs, managing energy supply and demand profiles, and integrating renewables in the buildings. In this chapter, energy demand items and their consumption profiles are first discussed for the buildings. This is followed by introductory information regarding the energy systems and devices used in the buildings.

References

[1] 2019 Global Status Report for Buildings and Construction, 2019. https://www. worldgbc.org/news-media/2019-global-status-report-buildings-and-construction. (Accessed 2 August 2020).

[2] Types of Residential Buildings and Their Site Selection, 2020. https://theconstructor. org/building/types-site-selection-residential-building/5995/#:~:text=A residential building is defined,or dining or both facilities. (Accessed 2 August 2020).

[3] European Union Energy Label, (n.d.). https://en.wikipedia.org/wiki/European_Union_ energy_label#:~:text=The energy efficiency of the,they choose between various models. (Accessed 2 August 2020).

[4] Energy Star, (n.d.). https://www.energystar.gov/. (Accessed 5 August 2020).

[5] V.S.K.V. Harish, A. Kumar, A review on modeling and simulation of building energy systems, Renew. Sustain. Energy Rev. 56 (2016) 1272–1292, https://doi.org/10.1016/j.rser.2015.12.040.

[6] Natural Resources Canada, Comprehensive Energy Use Database, (n.d.). https://oee.nrcan.gc.ca/corporate/statistics/neud/dpa/menus/trends/comprehensive_tables/list.cfm. (Accessed 9 August 2020).

[7] N. MacMackin, L. Miller, R. Carriveau, Modeling and disaggregating hourly effects of weather on sectoral electricity demand, Energy 188 (2019) 115956, https://doi.org/10.1016/J.ENERGY.2019.115956.

[8] D. Erdemir, I. Dincer, Potential use of thermal energy storage for shifting cooling and heating load to off-peak load: a case study for residential building in Canada, Energy Storage (2019) e125, https://doi.org/10.1002/est2.125.

[9] Center for Sustainable Systems, University of Michigan. Residential Building Factsheet, 2019. http://css.umich.edu/factsheets/residential-buildings-factsheet. (Accessed 9 August 2020).

[10] Y. Cengel, M. Boles, M. Kanoglu, Thermodynamics an Engineering Approach, ninth ed., 2019.

[11] I. Dincer, C. Acar, A review on clean energy solutions for better sustainability, Int. J. Energy Res. 39 (2015) 585–606, https://doi.org/10.1002/er.3329.

[12] California Opens "Diesel Alternative" Discussion in Microgrid Proceeding, 2020. https://microgridknowledge.com/california-microgrid-19-09-009/. (Accessed 19 September 2020).

System Analysis

4.1 Introduction

Further to what has been presented in the previous chapters, heat storage systems offer many advantages in terms of efficiency, cost, environmental impact, resource use and electricity grid load management. Although many parameters have significant impacts on the system performance, the following are considered critical for the performance, costs and savings of the heat storage system.

- System design parameters
- Physical conditions of the facility to be placed the heat storage system
- Environmental conditions
- Storage utilization strategy
- Energy source

To design optimum heat storage systems and gain the maximum benefits from them, the aforementioned criteria above should be taken into consideration.

For instance, for a hot water storage tank, the position and the diameter-to-height ratio of the tank significantly impact the stored energy amount and thermal stratification. Therefore, they should be set carefully by considering environmental conditions such as outdoor temperature, wind speed, and so on. For an ice storage system, it is critical to determine the ice production and storage methods, as they have a significant effect on the system's heat storage performance. Heat storage utilization strategy directly affects the initial investment costs and savings provided by the heat storage system. So, it should be carefully analyzed and applied to the system to reduce the system's payback period and increase the amount of savings. Consequently, to maximize the system benefit indexes and to minimize the initial investment cost and losses, modeling and analyzing the system is the first and essential issue. Then, simulation and optimization studies should be performed to determine the best possible system configuration.

One needs to remember that system analysis is an important step in dealing with heat storages. In this chapter, the methods used for analyzing and modeling the heat storage systems are introduced to the readers. As thermodynamics analysis based on the first and second laws of thermodynamics is essentially considered the prerequisite for the modeling and analysis of energy systems, first, the basics of the

Heat Storage Systems for Buildings. https://doi.org/10.1016/B978-0-12-823572-0.00003-5

115

thermodynamic analysis are presented. The balance equations for mass, energy, entropy, and exergy are introduced for the most used devices and processes in energy systems. Then, energy and exergy analyses of heat storage systems are presented with illustrative examples. It is followed by the basics of heat transfer and its applications of heat storage systems. Finally, the fundamentals of computational fluid mechanics (CFD) are introduced with a flow diagram.

4.2 Basic Thermodynamic Analysis

Thermodynamics is defined as the science of energy. However, today, thermodynamics approaches are used in many sectors and areas, from energy to environment and from sociology to economy. However, the first and also main purpose of the thermodynamic approaches is to better understand the energy systems and processes. A better understanding helps to improve the systems. The thermodynamically improved systems can provide more effective results in many ways such as less fuel consumption, power generation, the number of useful outputs, economy, environmental impacts, and so on.

To comprehend the fundamentals of thermodynamics analysis, it plays a significant role to understand the four laws of thermodynamics: the zeroth law, first law, second law, and third law of thermodynamics. While the first and second laws of thermodynamics cover methods to assess the systems with a set of rules, the zeroth and third laws of thermodynamics provide the guide to the first and second laws. The laws of thermodynamics are described by Dincer as the following [1]:

> The first and second laws of thermodynamics are recognized as governing laws like the constitutional laws for a state or country or institution which are known as the primary rules for regulating the functioning of a state or country or institution. When we look at the zeroth and third laws of thermodynamics, these are seen more as guiding policies for any state or country or institution. After this linkage, one may clearly understand that the first and second laws of thermodynamics are governing laws and the zeroth and third laws of thermodynamics are guiding laws depending on special/specific situations.

The zeroth law of thermodynamics can generally be defined with a supporting example, illustrated in Fig. 4.1. The zeroth law of thermodynamics states the equilibrium for the bodies/objects. If the body/object A is in equilibrium with the B and the body/object B is in equilibrium with the C, the body/object A is also in equilibrium with the C. Here, the critical question is what the equilibrium is or how the equilibrium conditions are to be defined. In the literature, the zeroth law of thermodynamics is often defined over the thermal equilibrium. Nonetheless, here, we should note that the equilibrium conditions should be defined through all parameters that have the potential to produce power, movement, flow, flux, or useful output. The zeroth law of thermodynamics provides information regarding the potential of the useful output or power generation between two objects/bodies/environments.

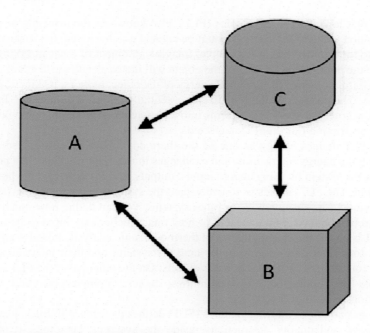

FIG. 4.1

An illustration of the zeroth law of the thermodynamics.

If two objects/bodies/environments are in the equilibrium conditions, there will be no power generation, movement, flux, flow, or useful output. Some equilibrium conditions may be listed as follows:

- Temperature
- Pressure
- Phase
- Concentration
- Electron
- Magnetic
- Chemical
- Force
- Vibration

   To ensure equilibrium conditions between the two systems/environments/bodies, the conditions listed before should be identical in each system/environment/body. When they are in equilibrium conditions, there will be no movement, flow, and flux between them. In other words, there is no potential to produce work or useful output with a system or process to be work between those systems. For instance, the batteries are able to produce power due to nonequilibrium electron conditions even if they are at the same temperature. It is possible to obtain a cooling effect with a water/ice mixture despite being at the same temperature.

The first law of thermodynamics (FLT), also known as the principle of energy conservation, defines that energy cannot be created nor destroyed, but it can change from one form to another. For example, imagine an object of mass m falling from above to the bottom: the velocity of the object will increase with the distance toward to downward. So, while the potential energy of the object reduces, its kinetic energy increases. The changes in the kinetic and potential energies are the same. In other words, the potential energy transforms into kinetic energy. Here, it should be noted that the air friction (air drag) is neglected.

The FLT defines energy as one of the thermodynamics properties and allows analyzing the energy interactions and exchanges in the systems. Thus, it is possible to assess the system by using energy inputs, outputs, and losses across the border of the system. The FLT enables us to understand the systems and interactions between the system and its surroundings or other systems. We can define how much useful output to be obtained from a system or how much fuel/energy input to be needed. The FLT is one of the essential items of thermodynamic analysis. It is also an essential tool for engineers to clearly understand the systems and their interactions with their surrounding environment or other objects. On the other hand, the FLT remains incapable of defining the irreversibilities, losses, inefficiencies, and quality destructions.

The second law of thermodynamics (SLT) defines the irreversibilities, losses, inefficiencies, and quality destructions inside the systems. Therefore, it is often described as a measure of the quality of energy. The SLT is related to both quantity and quality, while the FLT is only related to quantity. Therefore, the SLT gives detailed information regarding the systems. The difference between the FLT and SLT is expressed over the X-ray and tomography images. Although both are critical tools to examine the human body, the X-ray provides limited information according to the tomography. Here, while the FLT is the X-ray, the SLT is the tomography. To understand and evaluate the system clearly, the SLT is a powerful tool as it defines the maximum potential useful outputs to be obtained from the systems. The SLT is defined through entropy and exergy approaches.

The third law of thermodynamics (TLT) states that the entropy and entropy generation are zero at the 0 K that is called absolute zero. The TLT was developed by Walther Nernst, so it is called Nernst's theorem. In 1923, the TLT was modified by Gilbert N. Lewis and Merle Randall by defining the perfect crystalline substances. They presented that the perfect crystalline substances could occur at the absolute zero. In other words, it explains that there is no way to avoid entropy generation.

Thermodynamics is directly linked to energy, environment, and sustainability. Thermodynamics is located at the intersection of three, as illustrated in Fig. 4.2. Better thermodynamics performance can provide better performance indexes for energy, environment, and sustainability. Therefore, performing thermodynamic analysis is a critical tool to evaluate and optimize a system and system components. As expressed previously, thermodynamic analysis based on the FLT and SLT is a powerful tool for developing, improving, and optimizing energy systems.

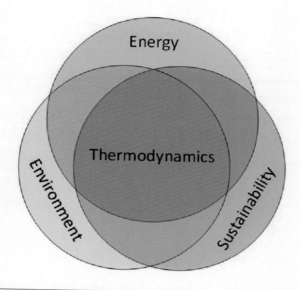

FIG. 4.2

An illustration of thermodynamics as an intersection of energy, environment, and sustainability.

Modified from Ref. [1].

To perform a thermodynamic analysis, the balance equations should be written for the system to be worked on. The key thermodynamic balance equations are written for mass, energy, entropy, and exergy in any thermodynamic system in either closed or open form.

Here, the thermodynamic system should be defined first. A thermodynamic system is defined as a mass or volume of matter, which is separated from its surroundings with an imaginary or physical border. The system and its border may be fixed in a stationary form or moving in a nonstationary form. While the FLT is taken into consideration the interactions that cross the system border, the changes inside the system are analyzed with the SLT, such as irreversibilities, generations, and destructions. A thermodynamic system is divided into two: the closed and open systems. In the closed system, there is no mass input or output into the system; in other words, the mass of the system is constant. In the open system, there are mass inlet(s) and outlet(s). Therefore, a volumetric region is the system.

To perform a thermodynamics analysis properly, the steps should be applied step-by-step, shown in Fig. 4.3. The boundary of the system is determined first. Then, all inputs and outputs (if any) are determined. This is followed by mass, energy, entropy, and exergy balances, respectively. Finally, the first and second law efficiencies are performed for performance assessment. To perform these steps located on the left side of Fig. 4.3, the steps shown on the right side should be followed. First, the properties of the materials and working fluids should be defined, and

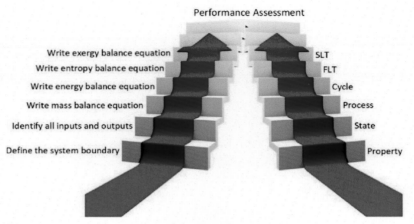

FIG. 4.3

A schematic view of the thermodynamics performance assessment.

then the states of the materials and working fluids should be determined according to their properties. Next, the type of processes is defined or assumed, such as isentropic, isobaric, isothermal, adiabatic, and so on. And these processes are combined to create the cycles. Finally, the FLT and SLT are applied to the cycles. This step-by-step approach is a systematic way to perform a thermodynamic analysis. It can be applied to all energy systems, from power generation systems to biological and chemical processes.

4.2.1 Thermodynamic Balance Equations for Common Components and Processes

Although all steps, shown in Fig. 4.3, are essential for a systematic thermodynamic analysis, writing the balance equations is the most important one. They should be written carefully for the systems or processes to be studied. The mass balance equation for an open system (a control volume) can be written in general form as

$$\sum \dot{m}_{in} - \sum \dot{m}_{out} = \frac{dm_{sys}}{dt} \tag{4.1}$$

where $\sum \dot{m}_{in}$ and $\sum \dot{m}_{out}$ are the total rate of input and output flows passing through the system boundary. $\frac{dm_{sys}}{dt}$ represents the mass change in the system. Under the steady condition, it equals to zero, $\frac{dm_{sys}}{dt} = 0$.

The energy balance equation in the general form is

$$\sum E_{in} - \sum E_{out} = \Delta E_{sys} \tag{4.2}$$

where $\sum E_{in}$, $\sum E_{out}$, and ΔE_{sys} represent the total energy input and output passing through the system boundary and energy change in the system, respectively.

The general form of the energy balance can be written in the rate form as

$$\sum \dot{E}_{in} - \sum \dot{E}_{out} = \frac{dE_{sys}}{dt} \tag{4.3}$$

Energy (E) for the closed system can be written as

$$E = U + KE + PE \tag{4.4}$$

Here, U denotes the internal energy. KE and PE are the kinetic and potential energies. The total energy of the closed system is the sum of its internal, kinetic, and potential energies, shown in Eq. (4.4). This can be written in rate form for the open systems (control volumes) as

$$\dot{E} = H + KE + PE \tag{4.5}$$

Here, H is the enthalpy. The kinetic and potential energy changes are generally neglected since the changes are quite negligible compared with the changes in internal energy or enthalpy in energy engineering applications.

The entropy balance equation can be written in the general form as

$$\sum S_{in} - \sum S_{out} + \sum S_{gen} = \Delta S_{sys} \tag{4.6}$$

where S_{in} and S_{out} represent the entropy values of the inputs and outputs passing through the system boundary. While S_{gen} is the entropy generation, ΔS_{sys} denotes the entropy change in the system. Eq. (4.6) can be written in the rate form as

$$\sum \dot{S}_{in} - \sum \dot{S}_{out} + \sum \dot{S}_{gen} = \frac{dS_{sys}}{dt} \tag{4.7}$$

Finally, the exergy balance equation can be written as

$$\sum Ex_{in} - \sum Ex_{out} - \sum Ex_{dest} = \Delta Ex_{sys} \tag{4.8}$$

where $\sum Ex_{in}$ and $\sum Ex_{out}$ denote the exergy inputs and outputs passing through the system boundaries. $\sum Ex_{dest}$ is the exergy destruction that denotes the irreversibilities in the system. The exergy balance equation can be written for the control volume in the rate form as

$$\sum \dot{Ex}_{in} - \sum \dot{Ex}_{out} - \sum \dot{Ex}_{dest} = \frac{dEx_{sys}}{dt} \tag{4.9}$$

Exergy terms in Eqs. (4.6) and (4.7) can be calculated as

$$Ex = (U_i - U_0) - T_0(S_i - S_0) + KE + PE \tag{4.10}$$

$$\dot{Ex} = (H_i - H_0) - T_0(S_i - S_0) + KE + PE \tag{4.11}$$

where "i" denotes the values for a state in the control volume and mass. "0" is the state at the reference conditions.

Also, here, it should be note that there is the following relation between exergy destruction and entropy generation.

$$Ex_{dest} = T_0 S_{gen} \tag{4.12}$$

$$\dot{Ex}_{dest} = T_0 \dot{S}_{gen} \tag{4.13}$$

The details of the thermodynamic balance equations and analysis are not limited to the aforementioned equations. Here, it is aimed to give a general perspective of thermodynamic analysis and balance equations. To learn about thermodynamic analysis and balance equations, please address the book *Thermodynamics: A Smart Approach* written by Ibrahim Dincer [1].

Almost all energy systems consist of similar components and processes. Here, thermodynamics balance equations of the common components or processes used in energy systems are introduced. The selected closed system examples are illustrated, and the balance equations for these systems are defined in Table 4.1. Of course, the examples of the components and processes given in Table 4.1 are not limited to them; however, the combination of the systems and processes can be used to analyze the different closed systems.

Thermodynamic balance equations for the selected open system (control volume) examples are presented in Table 4.2. The selected devices are commonly used in energy systems and heat storage systems in the buildings.

In the devices, systems, and processes in Tables 4.1 and 4.2, the potential and kinetic energy changes are neglected. To perform thermodynamic analysis, the conceptually correct assumptions, like in kinetic and potential energy changes, can be done to reduce the complexity of the analysis.

Illustrative Example 1. Power generation in a Rankine cycle

Consider an actual Rankine cycle, shown in Fig. 4.4a, steam enters the turbine 6 MPa and 450°C. The pressure of the condenser is 100 kPa. The temperature at the condenser outlet 90°C. The mass flow rate in the cycle is 15 kg/s.

a. Write thermodynamic balance equations for each component on the system.
b. Calculate the heat needed in the boiler.
c. Calculate the net power produced.
d. Calculate the energy and exergy efficiency of the cycle.

Solution:

In the thermodynamic analysis, it is first critical to draw the schematic diagram of the system studied. Also, for a thermodynamic cycle, its T-s diagram should be drawn along with it. The schematic diagram of the Rankine cycle and its T-s diagram are shown in Fig. 4.4.

Next, to reduce the complexity of the problem, the conceptually correct assumptions can be done. In the present example, the following assumptions are applied to the problem.

- The pressure losses in the boiler and condenser are neglected.
- The isentropic efficiency of the turbine is 90%.
- Reference conditions are assumed as 25°C and 100 kPa.
  a. Balance equations for each component on the Rankine cycle studied:
  Turbine:
  MBE: $\dot{m}_1 = \dot{m}_2$
  EBE: $\dot{m}_1 h_1 = \dot{m}_2 h_2 + \dot{W}_{turbine}$

Table 4.1 The Balance Equations for the Selected Closed Systems.

System or Process	Balance Equations
A rigid tank filled with a fluid subject to heat and work transfer	MBE: $m_1 = m_2 = m$: constant EBE: $m_1 u_1 + W_e + Q_{in} + W_{sh} = m_2 u_2 + Q_{loss}$ EnBE: $m_1 s_1 + \frac{Q_s}{T_s} + S_{gen} = m_2 s_2 + \frac{Q_{loss}}{T_0}$ ExBE: $m_1 ex_1 + W_e + Ex_{Q_s} + W_{sh} = m_2 ex_2 + Ex_{Q_{loss}} + Ex_{dest}$
A cooling bath with two solid blocks	MBE: $m_1 = m_2 = m$: constant for each material EBE: $m_{f,1}u_{f,1} + m_{sb1,1}u_{sb1,1} + m_{sb2,1}u_{sb2,1} = m_{f,2}u_{f,2} + m_{sb1,2}u_{sb1,2} + m_{sb2,2}u_{sb2,2}$ EnBE: $m_{f,1}s_{f,1} + m_{sb1,1}s_{sb1,1} + m_{sb2,1}s_{sb2,1} + S_{gen} = m_{f,2}s_{f,2} + m_{sb1,2}s_{sb1,2} + m_{sb2,2}s_{sb2,2}$ ExBE: $m_{f,1}ex_{f,1} + m_{sb1,1}ex_{sb1,1} + m_{sb2,1}ex_{sb2,1} = m_{f,2}ex_{f,2} + m_{sb1,2}ex_{sb1,2} + m_{sb2,2}ex_{sb2,2} + Ex_{dest}$

Continued

Table 4.1 The Balance Equations for the Selected Closed Systems.—*cont'd*

System or Process	Balance Equations
A compression process in a piston cylinder (the boundary work input)	MBE: $m_1 = m_2 = m$: constant EBE: $m_1 u_1 + W_b = m_2 u_2$ here $\quad\quad W_b = P\Delta v$ EnBE: $m_1 s_1 + S_{gen} = m_2 s_2$ ExBE: $m_1 ex_1 + W_b = m_2 ex_2 + Ex_{dest}$
An expansion process in a piston cylinder (the boundary work output)	MBE: $m_1 = m_2 = m$: constant EBE: $m_1 u_1 + Q_{in} + W_{sh} = m_2 u_2 + Q_{loss} + W_b$ EnBE: $m_1 s_1 + \dfrac{Q_{in}}{T_s} + S_{gen} = m_2 s_2 + \dfrac{Q_{loss}}{T_0} + W_b$ ExBE: $m_1 u_1 + Ex_{Q_{in}} + W_{sh} = m_2 u_2 + Ex_{Q_{loss}} + W_b + Ex_{dest}$

Table 4.2 The Balance Equations for the Selected Open Systems.

System or Process	Balance Equations
Pump	MBE: $\dot{m}_1 = \dot{m}_2 = \dot{m}$ EBE: $\dot{m}_1 h_1 + \dot{W}_{in} = \dot{m}_2 h_2$ EnBE: $\dot{m}_1 s_1 + \dot{S}_{gen} = \dot{m}_2 s_2$ ExBE: $\dot{m}_1 ex_1 + \dot{W}_{in} = \dot{ex}_2 h_2 + \dot{Ex}_{dest}$
Compressor	MBE: $\dot{m}_1 = \dot{m}_2 = \dot{m}$ EBE: $\dot{m}_1 h_1 + \dot{W}_{in} = \dot{m}_2 h_2$ EnBE: $\dot{m}_1 s_1 + \dot{S}_{gen} = \dot{m}_2 s_2$ ExBE: $\dot{m}_1 ex_1 + \dot{W}_{in} = \dot{ex}_2 h_2 + \dot{Ex}_{dest}$
Turbine	MBE: $\dot{m}_1 = \dot{m}_2 = \dot{m}$ EBE: $\dot{m}_1 h_1 = \dot{m}_2 h_2 + \dot{W}_{out}$ EnBE: $\dot{m}_1 s_1 + \dot{S}_{gen} = \dot{m}_2 s_2$ ExBE: $\dot{m}_1 ex_1 = \dot{ex}_2 h_2 + \dot{W}_{in} + \dot{Ex}_{dest}$
Expansion valve	MBE: $\dot{m}_1 = \dot{m}_2 = \dot{m}$ EBE: $\dot{m}_1 h_1 = \dot{m}_2 h_2$ EnBE: $\dot{m}_1 s_1 + \dot{S}_{gen} = \dot{m}_2 s_2$ ExBE: $\dot{m}_1 ex_1 = \dot{m}_2 ex_2 + \dot{Ex}_{dest}$

Continued

Table 4.2 The Balance Equations for the Selected Open Systems.—*cont'd*

System or Process	Balance Equations
Heat exchanger 	MBE: $m_1 = m_2$ *and* $m_3 = m_4$ EBE: $\dot{m}_1 h_1 + \dot{m}_3 h_3 = \dot{m}_2 h_2 + \dot{m}_4 h_4 + Q_{loss}$ EnBE: $\dot{m}_1 s_1 + \dot{m}_3 s_3 + \dot{S}_{gen} = \dot{m}_2 s_2 + \dot{m}_4 s_4 + \frac{Q_{loss}}{T_0}$ ExBE: $\dot{m}_1 ex_1 + \dot{m}_3 ex_3 = \dot{m}_2 ex_2 + \dot{m}_4 ex_4 + Ex_{Q_{loss}} + Ex_{dest}$
Mixing chamber 	MBE: $m_1 + m_2 = m_3 + m_4$ EBE: $\dot{m}_1 h_1 + \dot{m}_3 h_3 = \dot{m}_2 h_2 + \dot{m}_4 h_4$ EnBE: $\dot{m}_1 s_1 + \dot{m}_3 s_3 + \dot{S}_{gen} = \dot{m}_2 s_2 + \dot{m}_4 s_4$ ExBE: $\dot{m}_1 ex_1 + \dot{m}_3 ex_3 = \dot{m}_2 ex_2 + \dot{m}_4 ex_4 + Ex_{dest}$
Condenser 	MBE: $m_1 = m_2$ EBE: $\dot{m}_1 h_1 = \dot{m}_2 h_2 + \dot{Q}_{con}$ EnBE: $\dot{m}_1 s_1 + \dot{S}_{gen} = \dot{m}_2 s_2 + \frac{\dot{Q}_{con}}{T_b}$ ExBE: $\dot{m}_1 ex_1 = \dot{m}_2 ex_2 + \dot{Ex}_{Q_{con}} + \dot{Ex}_{dest}$
Evaporator 	MBE: $m_1 = m_2$ EBE: $\dot{m}_1 h_1 + \dot{Q}_{eva} = \dot{m}_2 h_2$ EnBE: $\dot{m}_1 s_1 + \dot{S}_{gen} + \frac{\dot{Q}_{eva}}{T_s} = \dot{m}_2 s_2$ ExBE: $\dot{m}_1 ex_1 + \dot{Ex}_{Q_{eva}} = \dot{m}_2 ex_2 + \dot{Ex}_{dest}$

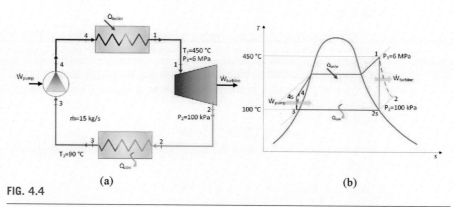

FIG. 4.4

(a) The schematic diagram and (b) T-s diagram of the steam Rankine cycle in illustrative example 1.

EnBE: $\dot{m}_1 s_1 + \dot{S}_{gen,turbine} = \dot{m}_2 s_2$

ExBE: $\dot{m}_1 ex_1 = \dot{ex}_2 h_2 + \dot{W}_{turbine} + \dot{Ex}_{dest,turbine}$

Condenser:

MBE: $\dot{m}_2 = \dot{m}_3$

EBE: $\dot{m}_2 h_2 = \dot{m}_3 h_3 + \dot{W}_{turbine}$

EnBE: $\dot{m}_2 s_2 + \dot{S}_{gen,con} = \dot{m}_3 s_3 + \dfrac{\dot{Q}_{con}}{T_b}$

ExBE: $\dot{m}_2 ex_2 = \dot{ex}_3 h_4 + \dot{Ex}_{\dot{Q}_{con}} + \dot{Ex}_{dest,con}$

Pump:

MBE: $\dot{m}_3 = \dot{m}_4$

EBE: $\dot{m}_3 h_3 + \dot{W}_{pump} = \dot{m}_4 h_4$

EnBE: $\dot{m}_3 s_3 + \dot{S}_{gen,pump} = \dot{m}_4 s_4$

ExBE: $\dot{m}_3 ex_3 + \dot{W}_{pump} = \dot{ex}_4 h_4 + \dot{Ex}_{dest,pump}$

Boiler:

MBE: $\dot{m}_4 = \dot{m}_1$

EBE: $\dot{m}_4 h_4 + \dot{Q}_{boiler} = \dot{m}_1 h_1$

EnBE: $\dot{m}_4 s_4 + \dfrac{\dot{Q}_{boiler}}{T_s} + \dot{S}_{gen,boiler} = \dot{m}_1 s_1$

ExBE: $\dot{m}_4 ex_4 + \dot{Ex}_{\dot{Q}_{boiler}} = \dot{ex}_1 h_1 + \dot{Ex}_{dest,boiler}$

b. To determine the heat given to the boiler, we need to use the EBE for the boiler. Here, we need to know that inlet and outlet states of the water circulated in the cycle. We already know the outlet state as it is turbine inlet. It is required to find the inlet state of boiler. The enthalpy of the stream entering the boiler is the same with the exiting from the pump.

$$w_{pump} = \nu_3 (P_4 - P_3) = 0.001010 \ x \ (6000 - 100) = 5.959 (kJ/kg)$$

where ν_3 is taken as the specific volume of the saturated liquid at T_1 and P_1.

$$w_{pump} = h_4 - h_3 \rightarrow 5.959 = (h_3 - 209.4) \rightarrow h_4 = 215.4$$

For the EBE of the boiler:

$$\dot{m}_4 h_4 + \dot{Q}_{boiler} = \dot{m}_1 h_1 \rightarrow \dot{Q}_{boiler} = 53883 \text{ kW}$$

c. In the cycle, while the steam turbine produces power, the pump consumes the power. The power generation in the turbine can be calculated by applying EBE as

$$\dot{m}_1 h_1 = \dot{m}_2 h_2 + \dot{W}_{turbine}$$

To calculate the turbine work, first, the isentropic work of the turbine should be calculated:

$$\dot{W}_{turbine,is} = \dot{m}(h_1 - h_{2s}) = 15 \times (3411 - 2675) = 736 \text{ kW}$$

Actual turbine work can be calculated:

$$\eta_{is} = \frac{\dot{W}_{turbine}}{\dot{W}_{turbine,is}} \rightarrow 0.90 = \frac{\dot{W}_{turbine,net}}{736} \rightarrow \dot{W}_{turbine} = 15520.32 \text{ kW}$$

d. The energy and exergy efficiencies of the Rankine cycle:
The overall energy efficiency of the Rankine cycle is obtained as

$$\eta_{overall} = \frac{(\dot{W}_{turbine,net} - \dot{W}_{pump})}{\dot{Q}_{boiler}}$$

The exergy efficiency of the Rankine cycle is obtained as

$$\psi_{overall} = \frac{(\dot{W}_{turbine,net} - \dot{W}_{pump})}{\dot{Ex}_{\dot{Q}_{boiler}}}$$

Illustrative Example 2.   An ideal refrigeration cycle
Consider a refrigeration system to keep the targeted volume at a temperature of −16°C. R134a is used as the refrigerant. R134a enters the compressor at 154.2 kPa. The flow rate of R134a is 0.32 kg/s. The refrigerant leaves the condenser at 24°C as a saturated liquid.

a. Write thermodynamic balance equations for each component in the cycle.
b. Find the rate of cooling provided by the system.
c. Calculate the energetic COP of the cycle.
d. Calculate the exergetic COP of the cycle.

Solution:
The schematic view of the refrigeration cycle and its P-h diagram are shown in Fig. 4.5.

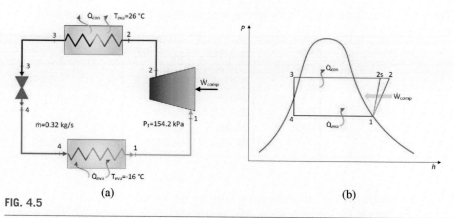

FIG. 4.5

(a) The schematic diagram and (b) P-h diagram of the refrigeration cycle in illustrative example 2.

The following assumptions are applied to reduce the complexity of the problem:

- The pressure losses in the evaporator, condenser, and expansion valve are neglected.
- The isentropic efficiency of the turbine is 70%.
- Reference conditions are assumed as 25°C and 100 kPa.

a) Balance equations for each component in the refrigeration cycle:

Compressor:
MBE: $\dot{m}_1 = \dot{m}_2$
EBE: $\dot{m}_1 h_1 + \dot{W}_{comp} = \dot{m}_2 h_2$
EnBE: $\dot{m}_1 s_1 + \dot{S}_{gen,comp} = \dot{m}_2 s_2$
ExBE: $\dot{m}_1 ex_1 + \dot{W}_{comp} = \dot{ex}_2 h_2 + \dot{Ex}_{dest,comp}$

Condenser:
MBE: $\dot{m}_2 = \dot{m}_3$
EBE: $\dot{m}_2 h_2 = \dot{m}_3 h_3 + \dot{W}_{turbine}$
EnBE: $\dot{m}_2 s_2 + \dot{S}_{gen,con} = \dot{m}_3 s_3 + \frac{\dot{Q}_{con}}{T_b}$
ExBE: $\dot{m}_2 ex_2 = \dot{ex}_3 h_4 + \dot{Ex}_{\dot{Q}_{con}} + \dot{Ex}_{dest,con}$

Expansion valve:
MBE: $\dot{m}_3 = \dot{m}_4$
EBE: $\dot{m}_3 h_3 = \dot{m}_4 h_4 \rightarrow h_3 = h_4$
EnBE: $\dot{m}_3 s_3 + \dot{S}_{gen,valve} = \dot{m}_4 s_4$
ExBE: $\dot{m}_3 ex_3 = \dot{ex}_4 h_4 + \dot{Ex}_{dest,valve}$

Evaporator:
MBE: $\dot{m}_4 = \dot{m}_1$
EBE: $\dot{m}_4 h_4 + \dot{Q}_{eva} = \dot{m}_1 h_1$
EnBE: $\dot{m}_4 s_4 + \frac{\dot{Q}_{eva}}{T_s} + \dot{S}_{gen,eva} = \dot{m}_1 s_1$
ExBE: $\dot{m}_4 ex_4 + \dot{Ex}_{\dot{Q}_{eva}} = \dot{ex}_1 h_1 + \dot{Ex}_{dest,eva}$

Table 4.3 The State Point Table of the Refrigeration Cycle in illustrative example 2.

State Point	Temperature (°C)	Pressure (kPa)	Specific Enthalpy (kJ/kg)	Specific Entropy (kJ/kgK)	Density (kg/m³)	Quality
1	−10	154.2	245.9	0.9633	7.581	
2	43.8	696	282.4	0.9768	30.73	
3	24	696	84.98	0.3194	1112	
4	−16	154.2	84.78	0.3357	6.785	0.26

b) Calculate the rate of cooling provided by the cycle:
To analyze the cycle, first the state points should be determined. State points of the cycle are shown in Table 4.3.
The rate of cooling provided by the cycle can be calculated by using EBE for the evaporator:

$$\dot{m}_4 h_4 + \dot{Q}_{eva} = \dot{m}_1 h_1 \rightarrow \dot{Q}_{eva} = 0.32 \ x \ (245.9 - 84.78) = 76.52 \ \text{kW}$$

c) Calculate the energetic COP of the cycle:

$$COP_{R,en} = \frac{\dot{Q}_{eva}}{\dot{W}_{comp}}$$

Here, the compressor work should first be found. It can be calculated from EBE for the compressor.

$$\dot{m}_1 h_1 + \dot{W}_{comp} = \dot{m}_2 h_2 \rightarrow \dot{W}_{comp} = 0.32 \ x \ (282.4 - 245.9) = 11.68 \ \text{kW}$$

and

$$COP_{R,en} = \frac{\dot{Q}_{eva}}{\dot{W}_{comp}} = \frac{76.52}{11.68} = 6.55$$

d) Calculate the exergetic COP of the cycle:

$$COP_{R,ex} = \frac{\dot{Ex}_{\dot{Q}_{eva}}}{\dot{W}_{comp}}$$

$$\dot{Ex}_{\dot{Q}_{eva}} = \dot{Q}_{eva}\left(\frac{T_0}{T_{eva}} - 1\right) = 76.52\left(\frac{298}{257} - 1\right) = 12.20 \ \text{kW}$$

$$COP_{R,ex} = \frac{12.20}{11.68} = 1.06$$

It should be noted here that the applications of thermodynamic analysis are not limited to these two examples. We have tried to give the basics of thermodynamic analysis for the systems, processes, and cycles. The introduced thermodynamic analysis approach can be applied to any problem. The same methodology has been used throughout the book in thermodynamic analyses.

4.3 Thermodynamic Balance Equations for Heat Storage Systems

Almost all heat storage systems consist of three operating stages: the charging, storing, and discharging periods, shown in Fig. 4.6. The duration of these stages

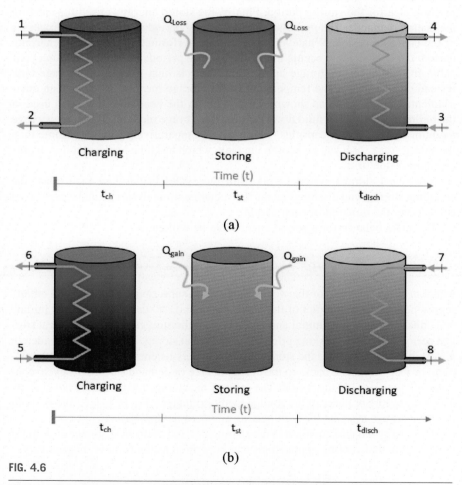

FIG. 4.6

The working principles of the heat storage systems for (a) heating purposes and (b) cooling purposes.

is determined according to the heat storage utilization strategy and availability of the energy source. In the energy loading period, one or more energy sources are converted to heat (if the energy source is not the heat) and transferred to a storage medium. If the heat storage is performed for heating purposes, the temperature of the storage medium increases or the storage medium is melted, or both (Fig. 4.6a). For heat storage for cooling purposes, the processes are the reverse from heating. Namely, the temperature of the storage medium decreases, or the storage medium is solidified, or both (Fig. 4.6b). The energy-charged storage medium is kept in a storage tank during the storing period until the energy is needed. The minimum losses during the storing period are expected. In certain heat storage systems, there may not be a storing period. For a heat storage system, the key solution is heat insulation. The well-insulated storage tanks, vessels, or volumes will provide the best performance for the storing period. Finally, the stored energy is used for meeting the energy demands. The processes in discharging period are the reverse processes of the charging period. Whatever processes are performed in the energy charging period, the opposite one occurs in the discharging period. These periods run cyclically. In the cold heat storage application, there is heat gain in the storage tank instead of heat loss as the temperature of the storage medium is lower than environmental temperature, as shown in Fig. 4.6b. In the heat storage system used in the buildings, a working fluid flows through the storage medium. Either the storage medium can be kept in capsules, which has a comparatively smaller volume than the storage tank, or it can be kept in the tank, which has tube banks that flow working fluid inside of it.

Since heat storage systems work with the cycles that follow each other, the charging, storing, and discharging, the thermodynamic analysis should be performed considering the duration of each period.

The mass balance for a period change can be written as

$$\sum_{t=0}^{t} \dot{m}_{inlet} \, t = \sum_{t=0}^{t} \dot{m}_{outlet} \, t \quad \text{and} \quad m_{sm,initial} = m_{sm,final} \tag{4.14}$$

where $\dot{m}$ is the flow rate of the working fluid, and subscripts of "inlet" and "outlet" represent the inlet and outlet of the heat storage unit. "t" denotes the time. "initial" and "final" represent the initial and final states of the storage medium in a heat storage period. When the working principles of the heat storage systems are considered, a working fluid flows into the storage tank, transfers its energy into the storage medium or vice versa, and backs to the energy source or demand sides. This cycle continues up to the end of the period. The mass of the storage medium does not change. The temperature or phase of the storage medium changes. Eq. (4.14) is valid for both the heat storage systems for both heating and cooling application.

The energy balance equation in the charging period of the heat storage system for both heating and cooling applications can be written as the follows, respectively:

$$\sum_{t=0}^{t_{ch}} (\dot{m}_1 (h_1 - h_2)) t = m_{sm} \left(u_{final} - u_{initial} \right) \tag{4.15}$$

$$\sum_{t=0}^{t_{ch}} (\dot{m}_1(h_5 - h_6))t = m_{sm}(u_{final} - u_{initial}) \tag{4.16}$$

The right side of Eqs. (4.15) and (4.16) is called the total charged energy. The charged energy changes the temperature or phase of the storage medium. The details of the sensible and latent heat storage systems will be discussed in the following sections.

The entropy balance equations for the charging period are

$$\left(\sum_{t=0}^{t_{ch}} \left(\dot{m}_1(s_1 - s_2) \right)t \right) + S_{gen} = m_{sm}(s_{final} - s_{initial}) \tag{4.17}$$

$$\left(\sum_{t=0}^{t_{ch}} \left(\dot{m}_1(s_5 - s_6) \right)t \right) + S_{gen} = m_{sm}(s_{final} - s_{initial}) \tag{4.18}$$

The exergy balances can be written as

$$\left(\sum_{t=0}^{t_{ch}} \left(\dot{m}_1(ex_1 - ex_2) \right)t \right) = m_{sm}(ex_{final} - ex_{initial}) + Ex_{dest} \tag{4.19}$$

$$\left(\sum_{t=0}^{t_{ch}} \left(\dot{m}_1(ex_5 - ex_6) \right)t \right) = m_{sm}(ex_{final} - ex_{initial}) + Ex_{dest} \tag{4.20}$$

The first terms on the right side of Eqs. (4.19) and (4.20) are called the total charged exergy into the storage medium.

In the storing period, there is no mass inlet and outlet in the storage medium or tank. The only interaction is the heat loss or heat gain. So, the mass balance can be written for the storing period as

$$m_{sm@initial} = m_{sm@final} \tag{4.21}$$

The energy balance equations:

$$m_{sm}u_{st@inital} = m_{sm}u_{st@final} + \sum_{t=0}^{t_{st}} \dot{Q}_{loss}t \tag{4.22}$$

$$m_{sm}u_{st@inital} + \sum_{t=0}^{t_{st}} \dot{Q}_{gain}t = m_{sm}s_{st@final} \tag{4.23}$$

The entropy balance equation:

$$m_{sm}s_{st@inital} + S_{gen} = m_{sm}s_{st@final} + \sum_{t=0}^{t_{st}} \frac{\dot{Q}_{loss}}{T_0}t \tag{4.24}$$

$$m_{sm}s_{st@inital} + \left(\sum_{t=0}^{t_{st}} \frac{\dot{Q}_{gain}}{T_0}t \right) + S_{gen} = m_{sm}s_{st@final} \tag{4.25}$$

The exergy balance equations:

$$m_{sm}ex_{st@inital} = m_{sm}s_{st@final} + \sum_{t=0}^{t_{st}}\left[\left(1 - \frac{T_0}{T}\right)\dot{Q}_{loss}\right]t + Ex_{dest} \qquad (4.26)$$

$$m_{sm}s_{st@inital} + \sum_{t=0}^{t_{st}}\left[\left(1 - \frac{T_0}{T}\right)\dot{Q}_{gain}\right]t = m_{sm}s_{st@final} + Ex_{dest} \qquad (4.27)$$

In the discharging period, the mass balance equations are same with Eq. (4.14). The energy balance equation for the discharging period:

$$\sum_{t=0}^{t_{ch}}\left(\dot{m}_3(h_3 - h_4)\right)t = m_{sm}\left(u_{disch@initial} - u_{disch@final}\right) \qquad (4.28)$$

$$\sum_{t=0}^{t_{ch}}\left(\dot{m}_7(h_7 - h_8)\right)t = m_{sm}\left(u_{disch@initial} - u_{disch@final}\right) \qquad (4.29)$$

Here, the right side of Eqs. (4.28) and (4.29) is called the total discharged energy or useful energy obtained by the heat storage system.

The entropy balance equations in the discharging period:

$$\sum_{t=0}^{t_{ch}}\left(\dot{m}_3(s_3 - s_4)\right)t + S_{gen} = m_{sm}\left(s_{disch@initial} - s_{disch@final}\right) \qquad (4.30)$$

$$\sum_{t=0}^{t_{ch}}\left(\dot{m}_7(s_7 - s_8)\right)t + S_{gen} = m_{sm}\left(s_{disch@initial} - s_{disch@final}\right) \qquad (4.31)$$

The exergy balance equations in the discharging period:

$$\left(\sum_{t=0}^{t_{ch}}\left(\dot{m}_3(ex_3 - ex_4)\right)t\right) = m_{sm}\left(ex_{disch@initial} - ex_{disch@final}\right) + Ex_{dest} \qquad (4.32)$$

$$\left(\sum_{t=0}^{t_{ch}}\left(\dot{m}_7(ex_7 - ex_8)\right)t\right) = m_{sm}\left(ex_{disch@initial} - ex_{disch@final}\right) + Ex_{dest} \qquad (4.33)$$

4.4 Energy and Exergy Analyses of Sensible Heat Storage Systems

Here, energy and exergy analyses for the sensible heat storage systems are introduced. The expressions are given through an illustrative example.

Consider a hot water storage tank (like in Fig. 4.4a). The heat transfer fluid (HTF), also called working fluid, circulates between the hot water tank and energy source (solar, burner, furnace, etc.). The HTF at the high temperature flows to the storage tank. The heat carried by HTF transfers to the water in the tank, and the temperature of the water in the tank increases. This process is the charging period. The hot water is kept in the storage until it is needed, and this period is called the storing period. In the discharging period, the stored hot water is used to meet the demand.

The overall energy balance for the hot water storage tank can be written as

(*Energy charged*) − [(*Energy discharged*) + (*Energy loss*)] = (*Energy accumulation*)

and

$$Q_{ch} - [Q_{disch} + Q_{loss}] = \Delta E_{sm} \tag{4.34}$$

While Q_{ch} represents the total charged energy, Q_{disch} denotes the total discharged energy. Q_{loss} is the total heat loss from tank to the surrounding. ΔE_{sm} is the internal energy change in the storage medium. If the heat storage system undergoes a complete cycle, the initial and final temperature of the storage medium will be equal. So, the internal energy change in the storage medium after a complete cycle is zero ($\Delta E_{sm} = 0$).

Q_{ch}, Q_{disch}, and ΔE_{sm} can be calculated as the follows:

$$Q_{ch} = \int_{t=0}^{t_{ch}} \dot{m}_1 (h_1 - h_2) dt \quad \text{(J, kJ or kWh)} \tag{4.35}$$

$$Q_{disch} = \int_{t=0}^{t_{disch}} \dot{m}_3 (h_3 - h_3) dt \quad \text{(J, kJ or kWh)} \tag{4.36}$$

$$\Delta E_{sm} = m_{sm} \left(u_{final} - u_{initial} \right) \tag{4.37}$$

Also, the energy balance equation for the charging and discharging periods for a sensible heat storage system can be written as

$$\sum_{t=0}^{t_{ch}} \left(\dot{m}_1 (h_1 - h_2) \right) t = m_{sm} \left(u_{final} - u_{initial} \right) \tag{4.38}$$

$$\sum_{t=0}^{t_{ch}} \left(\dot{m}_3 (h_3 - h_4) \right) t = m_{sm} \left(u_{initial} - u_{final} \right) \tag{4.39}$$

The overall exergy balance for the hot water storage tank can be written as

(*Exergy charged*) − [*Exergy discharged* + *Exergy loss due to heat losses*]
− (*Exergy destruction*) = *Exergy accumulation*

and

$$Ex_{ch} - \left[Ex_{disch} + Ex_{Q_{loss}} \right] - Ex_{dest} = \Delta Ex_{sm} \tag{4.40}$$

where Ex_{ch} and Ex_{disch} are the charged and discharged exergies, respectively. $Ex_{Q_{loss}}$ is the exergy loss due to heat loss from the tank to surrounding. Ex_{dest} is the exergy destruction due to irreversibilities in the system. ΔEx_{sm} is the exergy change in the system between the initial and final states of the hot water tank.

Ex_{ch}, Ex_{disch}, and ΔEx_{sm} can be calculated with

$$Ex_{ch} = \int_{t=0}^{t_{ch}} \dot{m}_1 (ex_1 - ex_2) dt \quad \text{(J, kJ or kWh)} \tag{4.41}$$

$$Ex_{disch} = \int_{t=0}^{t_{ch}} \dot{m}_3 (ex_3 - ex_3) dt \quad \text{(J, kJ or kWh)} \tag{4.42}$$

$$\Delta Ex_{sm} = m_{sm} \left(ex_{final} - ex_{initial} \right) \tag{4.43}$$

The specific exergy values can be calculated with

$$ex = (h_i - h_0) - T_0(s_i - s_0) \tag{4.44}$$

Note that the kinetic and potential energy changes are neglected in Eq. (4.44).

Also, the exergy balance equation for the charging and discharging periods for a sensible heat storage system can be written as

$$\sum_{t=0}^{t_{ch}} \left(\dot{m}_1(ex_1 - ex_2) \right) t - Ex_{dest} = m_{sm} \left(ex_{final} - ex_{initial} \right) \tag{4.45}$$

$$\sum_{t=0}^{t_{ch}} \left(\dot{m}_3(ex_3 - ex_4) \right) t - Ex_{dest} = m_{sm} \left(ex_{initial} - ex_{final} \right) \tag{4.46}$$

The aforementioned balance equations given are applied to calculate the energetic and exergetic efficiencies. Now, the energy and exergy efficiencies for each period and overall system will be introduced.

Energy efficiency for a sensible heat storage system can be defined as

$$\eta = \frac{Net\ discharged\ energy\ (net\ energy\ output)}{Charged\ energy\ (energy\ input)} \tag{4.47}$$

Therefore, the overall energy efficiency:

$$\eta_{overall} = \frac{Q_{disch}}{Q_{ch}} = \frac{\sum\limits_{t=0}^{t_{ch}} \left(\dot{m}_1(h_1 - h_2) \right) t}{\sum\limits_{t=0}^{t_{disch}} \left(\dot{m}_3(h_3 - h_4) \right) t} \tag{4.48}$$

Energy efficiencies for the operating periods, which are charging, storing, and discharging periods, can be written as

$$\eta_{ch} = \frac{m_{sm} \left(u_{ch@final} - u_{ch@initial} \right)}{\sum\limits_{t=0}^{t_{ch}} \left(\dot{m}_1(h_1 - h_2) \right) t} \tag{4.49}$$

$$\eta_{st} = \frac{m_{sm} \left(u_{st@final} - u_{st@initial} \right) - Q_{loss}}{m_{sm} \left(u_{st@final} - u_{st@initial} \right)} \tag{4.50}$$

$$\eta_{disch} = \frac{\sum\limits_{t=0}^{t_{ch}} \left(\dot{m}_3(h_3 - h_4) \right) t}{m_{sm} \left(u_{ch@initial} - u_{ch@final} \right)} \tag{4.51}$$

Exergy efficiency for a sensible heat storage system can be defined as

$$\psi = \frac{Net\ discharged\ exergy\ (net\ exergy\ output)}{Charged\ exergy\ (exergy\ input)} \tag{4.52}$$

Therefore, the overall energy efficiency:

$$\psi_{overall} = \frac{Ex_{disch}}{Ex_{ch}} = \frac{\sum\limits_{t=0}^{t_{disch}} \left(\dot{m}_1 (ex_1 - ex_2) \right) t}{\sum\limits_{t=0}^{t_{ch}} \left(\dot{m}_3 (ex_3 - ex_4) \right) t} \qquad (4.53)$$

Energy efficiencies for the operating periods, which are charging, storing, and discharging periods, can be written as

$$\psi_{ch} = \frac{m_{sm} \left(ex_{ch@final} - ex_{ch@initial} \right)}{\sum\limits_{t=0}^{t_{ch}} \left(\dot{m}_1 (ex_1 - ex_2) \right) t} \qquad (4.54)$$

$$\psi_{st} = \frac{m_{sm} \left(ex_{st@final} - ex_{st@initial} \right) - Q_{loss}}{m_{sm} \left(ex_{st@final} - ex_{st@initial} \right)} \qquad (4.55)$$

$$\psi_{disch} = \frac{\sum\limits_{t=0}^{t_{ch}} \left(\dot{m}_3 (ex_3 - ex_4) \right) t}{m_{sm} \left(ex_{ch@initial} - ex_{ch@final} \right)} \qquad (4.56)$$

Illustrative Example 3.   A mantled hot water storage tank

Consider 500 L mantled hot water storage tank, commonly used in the solar domestic hot water systems, shown in Fig. 4.7. In the charging period, the temperatures at the inlet and outlet of the mantle are 95°C and 50°C, respectively. The charging period takes 6 h. Then, hot water produced is kept for 3 h in the tank. Finally, the stored water is used until the water temperature becomes 20°C, which is the main line inlet temperature. In the beginning, the water temperature in the tank is 20°C, and there is no thermal stratification. The heat transfer coefficient between the tank wall and surrounding is 100 W/m²K. The flow rate in the mantle is 0.01 kg/s. Evaluate the hot water storage process based on FLT and SLT.

Solution: The assumptions applied to the problem are defined as follows:

- Mantle inlet and outlet temperatures are constant during the charging period.
- Heat loss is only considered in the storing period.
- Pressure losses in the system are neglected.
- There is no thermal stratification in all operating periods.
- Reference conditions are 25°C and 100 kPa.

Charging period:

MBE: $\dot{m}_1 = \dot{m}_2 = \dot{m}$ and $m_{hws@initial} = m_{hws@final}$

EBE: $\sum\limits_{t=0}^{t_{ch}} \left(\dot{m}_1 h_1 - \dot{m}_2 h_2 \right) t = m_{hws} \left(u_{final} - u_{initial} \right)$ or

$$\sum\limits_{t=0}^{t_{ch}} \left(\dot{m}_1 h_1 - \dot{m}_2 h_2 \right) t = m_{hws} c_p \left(T_{final} - T_{inital} \right)$$

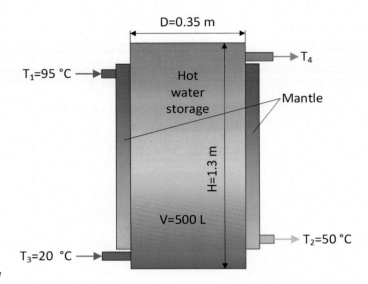

FIG. 4.7

A mantled hot water storage tank widely used in solar domestic hot water systems.

EnBE: $\sum_{t=0}^{t_{ch}}(\dot{m}_1 s_1 + \dot{S}_{gen,ch} - \dot{m}_2 s_2)t = m_{hws}(s_{final} - s_{initial})$

ExBE: $\sum_{t=0}^{t_{ch}}(\dot{m}_1 ex_1 - \dot{m}_2 ex_2 - \dot{Ex}_{dest,ch})t = m_{hws}(ex_{final} - ex_{initial})$

The amount of the charged energy can be obtained from the EBE. Here, $h_1 = 398.1$ kJ/kg and $h_2 = 209.4$ kJ/kg.

$$\dot{Q}_{ch} = \dot{m}(h_1 - h_2) = 0.03 * (398.1 - 209.4) = 5.66 \text{ kW}$$

$$Q_{ch} = \dot{m}(h_1 - h_2)t_{ch} = 0.03 * (398.1 - 209.4) * (6 * 3600) = 122267 \text{ kJ}$$

The charged energy will change the water temperature in the storage tank, as shown in the EBE. The temperature of the stored water at the end of charging period can be obtained as

$$Q_{ch} = m_{hws} c_p (T_{final} - T_{inital}) = 976.3 * 0.5 * 4.192 * (T_{final} - 20) \rightarrow T_{final} = 79.75°\text{C}$$

Here, the density and specific heat of the water are taken according to the average temperature of the mantle inlet and outlet. $T_{ave,ch} = \frac{(T_1 + T_2)}{2} = 72.5°C$. There is no significant difference between $T_{final} = 79.75°C$ and $T_{ave,ch} = 72.5°C$ affecting the properties of the water. It can be validated by solving the same equation with the updated properties at the temperature of $T_{final} = 79.75°C$.

$$Q_{ch} = m_{hws}c_p\left(T_{final,new} - T_{inital}\right) = 971.9 * 0.5 * 4.197 * \left(T_{final,new} - 20\right)$$

$$T_{final,new} = 79.95°C$$

The difference between T_{final} and $T_{final,new}$ is only $0.20°C$. To increase accuracy, the same equation can be solved with updating properties of the working fluid according to $T_{final,new}$. If one more step is applied with the properties at the temperature of $79.95°C$:

$$Q_{ch} = m_{hws}c_p\left(T_{final,new,2} - T_{inital}\right) = 971.8 * 0.5 * 4.197 * \left(T_{final,new,2} - 20\right)$$

$$T_{final,new,2} = 79.96°C$$

Now, the difference is only $0.01°C$. The accuracy of the solution should be determined according to the accuracy of the thermodynamic tables that obtaining the properties.

To determine the entropy generation during the charging period, the EnBE equation should be solved.

$$\frac{Q_{ch}}{T_{mantle,ave}} + S_{gen,ch} = m_{hws}\left(s_{final} - s_{initial}\right)$$

$$\frac{122267}{(72.5 + 273.15)} + S_{gen,ch} = 971.8 * 0.5 * (1.073 - 0.2965)$$

$$S_{gen,ch} = 23.41 \text{ kJ}$$

ExBE is solved to obtain the exergy destruction during the charging period.

$$Ex_{Q_{ch}} - Ex_{dest,ch} = m_{hws}\left(ex_{final} - ex_{initial}\right)$$

$$Ex_{Q_{ch}} = Q_{ch}\left(1 - \frac{T_0}{T_{mantle,ave}}\right) = 122267\left(1 - \frac{25 + 273.15}{72.5 + 273.15}\right) = 16802 \text{ kJ}$$

$$ex_{initial} = \left(u_{initial} - u_0\right) - T_0\left(s_{initial} - s_0\right) = 19.63 \text{ kJ/kg}$$

$$ex_{final} = \left(u_{final} - u_0\right) - T_0\left(s_{final} - s_0\right) = 0.17 \text{ kJ/kg}$$

$$Ex_{dest,ch} = 7349 \text{ kJ}$$

Storing period:
During the storing period, there is no mass inlet and outlet in the hot water storage tank. The only interaction is the heat loss, which is $100 \text{ W/m}^2\text{K}$.

MBE: $m_{hws@initial} = m_{hws@final} = m_{hws} = \rho * V_{tank}$

EBE: $\displaystyle\sum_{t=0}^{t_{st}}\left(-\dot{Q}_{loss}\right)t = m_{hws}c_p\left(T_{final,st} - T_{inital,st}\right)$

EnBE: $\displaystyle\sum_{t=0}^{t_{st}}(\frac{-\dot{Q}_{loss}}{T_0} + \dot{S}_{gen.st})t = m_{hws}\left(s_{final,st} - s_{initial,st}\right)$

ExBE: $\sum_{t=0}^{t_{st}} (\dot{Ex}_{\dot{Q}_{loss}} - \dot{Ex}_{dest.st})t = m_{hws}(ex_{final.st} - ex_{initial,st})$

The heat loss causes the temperature decrease in the water stored in the tank. The duration of the storing period is $t_{st} = 3$ h. Final temperature of the stored water at the end of storing period can be obtained from the EBE. First, it is required to calculate $\dot{Q}_{loss}$:

$$\dot{Q}_{loss} = h * A_{tank,sur} = 100 * 1.621 = 0.16 \text{ kW}$$

$$A_{tank,sur} = \pi * D * L + 2 * \left(\pi * \frac{D^2}{4}\right) = 1.621 \text{ m}^2$$

It is required to determine the density and specific heat of the water for applying the EBE. To determine them, the temperature of the water should first be assumed. As the first assumption, the temperature of the water is taken as 79.96°C. Also, at the beginning of the storing period, the temperature of water is 79.96°C, which is the final temperature of the charging period.

$$-(0.16 * 3 * 3600) = 971.8 * 0.5 * (T_{final,st} - 79.96)$$

$$T_{final,st} = 79.12°C$$

If the properties of the water are updated according to $T_{final.st} = 79.12°C$:

$$T_{final,st,new} = 79.10°C$$

At the end of the storing period, the water temperature decreases approximately 1°C.

As we determine the final temperature of the water in the storing period, now, we can calculate the entropy generation and exergy destruction during the storing period. From the EnBE:

$$-\frac{Q_{loss}}{T_0} + S_{gen.st} = m_{hws}(s_{final,st} - s_{initial,st})$$

$$-\frac{1751}{25 + 273.15} + S_{gen.st} = 971.8 * 0.5 * (1.065 - 1.075)$$

$$S_{gen.st} = 0.9075 \text{ kJ}$$

ExBE:

$$Ex_{Q_{loss}} - Ex_{dest.st} = m_{hws}(ex_{final.st} - ex_{initial,st})$$

$$Ex_{Q_{loss}} = Q_{loss}\left(1 - \frac{T_0}{T_{ave,st}}\right) = 1751\left(1 - \frac{25 + 273.15}{\left(\frac{79.96 + 79.10}{2}\right) + 273.15}\right) = 207.7\text{kJ}$$

$$ex_{initial,st} = (u_{initial,st} - u_0) - T_0(s_{initial,st} - s_0) = 19.63 \text{ kJ/kg}$$

$$ex_{final,st} = (u_{final,st} - u_0) - T_0(s_{final,st} - s_0) = 18.36 \text{ kJ/kg}$$

$$Ex_{dest.st} = 541.3 \text{ kJ}$$

Discharging period:

In the discharging period, the water provided by mains (T_3) enters the tank, and the hot water is discharged (T_4). There is no flow in the mantle. The balance equations for the discharging period:

MBE: $\dot{m}_3 = \dot{m}_4 = \dot{m}$

EBE: $\displaystyle\sum_{t=0}^{t_{disch}} \left(\dot{m}_3 h_3 - \dot{m}_4 h_4 \right) t = m_{hws} \left(u_{final,disch} - u_{initial,disch} \right)$ or

$$\sum_{t=0}^{t_{disch}} \left(\dot{m}_3 h_3 - \dot{m}_4 h_4 \right) t = m_{hws} c_p \left(T_{final,disch} - T_{inital,disch} \right)$$

In the discharging period, since all stored hot water is used, we know the initial and final temperature of the water. While the initial temperature of the water is $T_{inital,disch} = 79.10°C$, which is the final temperature of the water at the end of storing period, the final temperature of the water is $T_{final,disch} = 20°C$, which is the temperature of the mains. The total amount of the discharged energy is obtained as

$$Q_{disch} = m_{hws} c_p \left(T_{final,disch} - T_{inital,disch} \right)$$

The properties of the water can be obtained by using the average temperature, which is $\left(T_{final,disch} + T_{inital,disch} \right)/2$.

$$Q_{disch} = 988.2 * 0.5 * 4.181 * (79.10 - 20) = 122105 \text{ kJ}$$

By using EBE equation, we can observe two parameters: the duration of the discharging (t_{disch}) or flow rate ($\dot{m}$). To calculate one of them, it is required to know one of them. Here, we assumed that the hot water is discharged with the flow rate of $\dot{V} = 10 \text{ L/min}$.

$$\dot{m}(h_3 - h_4)t_{disch} = Q_{disch} \rightarrow \dot{V}\rho(h_3 - h_4)t_{disch} = Q_{disch}$$

$$10\left(\frac{L}{min}\right) * \frac{1}{60000} * 998.2 * (84.01 - 331.3)t_{disch} = -Q_{disch}$$

$$t_{disch} = 50 \text{ min}$$

To calculate the entropy generation in the discharging period, the EnBE should be solved:

EnBE:

$$\sum_{t=0}^{t_{disch}} \left[\dot{m}_3 (s_3 - s_4)t \right] + S_{gen,disch} = m_{hws} \left(s_{final,disch} - s_{initial,disch} \right)$$

Here, there is a specific situation in the discharging period. Since they have the equal temperatures:

$$s_3 = s_{final,disch} \text{ and } s_4 = s_{initial,disch}$$

Also, all water is replaced with the water coming from mains:

$$\dot{m}_3 * t_{disch} = m_{hws}$$

the entropy generation in the discharging period is calculated to be

$$S_{gen,disch} = 0.24 \text{ kJ}$$

To calculate the exergy destruction in the discharging period, the ExBE is solved. **ExBE**:

$$\sum_{t=0}^{t_{disch}} \left[\dot{m}_3 (ex_3 - ex_4) t \right] - Ex_{dest,disch} = m_{hws} \left(ex_{final,disch} - ex_{initial,disch} \right)$$

Similar with the entropy generation, it is expected that the exergy destruction should be lower due to only the stored water is discharged. The exergy destruction is found to be

$$Ex_{dest,disch} = 5.86 \text{ kJ}$$

Overall energy and exergy efficiencies for the storage tank:

$$\eta_{ove} = \frac{Total\ discharged\ energy}{Total\ charged\ energy} = \frac{Q_{disch}}{Q_{ch}} = \frac{122105}{122267} = 0.99$$

$$\psi_{ove} = \frac{Total\ discharged\ exergy}{Total\ charged\ exergy} = \frac{Q_{disch}}{Q_{ch}} = \frac{9290}{16802} = 0.55$$

Discussion: Hot water storage tanks are one of the most common heat storage applications. The method used in the present examples is almost same for all sensible heat storage systems. Depending on the application, there may be special conditions or assumptions. As can be seen from overall energy and exergy efficiencies, while energy efficiency is almost 100%, the exergy efficiency is around 55%. Although the heat storage studied in the example seems ideal according to energy efficiency, exergy efficiency shows how the heat storage process is far away from the ideal, with the value of 55%. In the discharging period, since the stored hot water is only discharged (there no any interaction crossing the system boundary), there is no entropy generation and exergy destruction. The highest exergy destruction is seen in the charging period.

4.5 Energy and Exergy Analyses of Latent Heat Storage Systems

In latent heat storage, unlike sensible heat storage, heat storage mainly occurs with phase change. The storage medium is called the phase change material (PCM). Also, before and after the phase change temperature, sensible heat storage can take place. The energy change in the PCM is shown in Fig. 4.8. The amount of the stored energy as latent heat is quite higher than the sensible one. In the thermodynamic analyses,

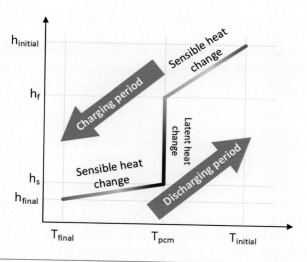

FIG. 4.8

The energy change in the ice storage system during the charging and discharging periods.

the phase change should be included the calculation. In sensible heat storage systems, heat is charged and discharged in the storage medium with internal energy change due to temperature change. Therefore, the energy change in the storage medium is defined as

$$\Delta E_{sm} = m_{sm}\left(u_{final} - u_{inital}\right) \quad \text{or}$$
$$\Delta E_{sm} = m_{sm}c\left(T_{final} - T_{inital}\right) \tag{4.57}$$

Similarly, the exergy change in the storage medium is

$$\Delta Ex_{sm} = m_{sm}\left(ex_{final} - ex_{inital}\right) \tag{4.58}$$

In latent heat storage, unlike sensible heat storage, heat storage mainly occurs with phase change. The storage medium is called the PCM. Also, before and after the phase change temperature, sensible heat storage can take place. The energy change in the PCM is shown in Fig. 4.8. The amount of the stored energy as latent heat is relatively higher than the sensible one. In the thermodynamic analyses, the phase change should be involved in the calculation. The energy and exergy changes in the storage medium can be written as

$$\Delta E_{sm} = m_{sm}\left(u_{inital} - u_{pcm,f}\right) + \left(h_{sf}\right) + c_s\left(T_{pcm,s} - T_{final}\right)$$
$$\Delta E_{sm} = m_{sm}\left[\left(c_f\left(u_{inital} - T_{pcm}\right)\right) + \left(h_{sf}\right) + \left(c_s\left(T_{pcm} - T_{final}\right)\right)\right] \tag{4.59}$$

where s and f subscripts denote solid and fluid phases of the PCM. h_{sf} is the enthalpy of phase change.

The exergy change in the PCM can be defined as

$$\Delta Ex_{sm} = m_{sm}\left(ex_{final} - ex_{initial}\right) \tag{4.60}$$

where final represents the exergy of the final state of PCM and initial represents the exergy of the initial state of PCM.

When the PCM consists of a mixture of solid and liquid phases. The energy change in the PCM can be defined as

$$\Delta E_{sm} = m_{sm}\left[(1-F)\left(h_{final,s} - h_{initial}\right) + F\left(h_{final,f} - h_{initial}\right)\right] \tag{4.61}$$

Here, F is the liquid fraction in the mixture. $h_{final,s}$ and $h_{final,f}$ are the internal energies of the solid and liquid phases of the PCM.

Here, the equations for energy and exergy analyses are introduced for the latent heat storage system. Ice thermal energy storage systems also called ice storage systems are one of the most common heat storage methods of latent heat and cold capacity storage. Here, the equations for energy and exergy analyses of the latent heat storage are introduced over an ice storage system. Fig. 4.4b demonstrates the working principles of the ice storage system. An HTF flows through the ice capsules or inside the tubes. For the details of the ice storage system, see Section 2.9. The energy change in the storage medium (water/ice) is shown in Fig. 4.8.

Consider an ice storage tank, as shown in Fig. 4.4b. During the charging period, the HTF is pumped to the tank at the temperature below 0°C, and it extracts the heat from the water: thus the ice is produced. The produced ice is kept in a well-insulated tank during the storing period. In the discharging period, the HTF is pumped to the tank at the temperature above 0°C. The temperature of the HTF decreases, as it losses heat and ice melts. The cooling demands are then met by the ice stored.

The overall energy balance for the ice storage tank can be written as

$$(- Charged\ cold\ capacity) + (Discharged\ cold\ capacity) + (Heat\ gain)$$
$$= (Energy\ accumulation)$$

and

$$-Q_{ch} + Q_{disch} + Q_{gain} = \Delta E_{sm} \tag{4.62}$$

Q_{ch} and Q_{disch} are the total charged and discharged cold capacities. While the sign of the charged cold capacity is the minus, it is positive for the discharged cold capacity. This occurs because of the direction of the heat transfer. In the charging period, the direction of the heat transfer is from the storage tank to HTF. In the discharging period, it is reverse. The sign notation is opposite from Eq. (4.62) since the heat fluxes are reverse according to cold storage.

Q_{ch} and Q_{disch} can be written by considering Fig. 4.4b as

$$Q_{ch} = \int_{t=0}^{t_{ch}} \dot{m}_5(h_5 - h_6)dt \quad (J,\ kJ)$$
$$Q_{ch} = \sum_{t=0}^{t_{ch}} \dot{m}_5(h_5 - h_6)\,t \quad (J,\ kJ\ or\ kW) \tag{4.63}$$

$$Q_{disch} = \int_{t=0}^{t_{disch}} \dot{m}_7(h_7 - h_8)dt \quad \text{(J, kJ)}$$

$$Q_{disch} = \sum_{t=0}^{t_{disch}} \dot{m}_8(h_7 - h_8)t \quad \text{(J, kJ or kW)}$$

(4.64)

Here, Q_{ch} and Q_{disch} are equal to the energy change in the PCMs used as the storage medium (ΔE_{sm}). ΔE_{sm} consists of the stages, shown in Fig. 4.8. The equation of ΔE_{sm} is given in Eq. (4.59).

1. Sensible heat changes in the water: temperature decrease
2. Latent heat change: solidification of the water
3. Sensible heat changes in the ice: temperature increase

The energy balance equation for the charging, storing, and discharging periods of the latent heat storage system can be written as

$$\sum_{t=0}^{t_{ch}} \left(\dot{m}_5 h_5 - \dot{m}_6 h_6 \right) t = m_{sm} \left[\left(c_f \left(T_{t=0} - T_{pcm} \right) \right) + \left[(1-F)\left(h_{t=0,s} - h_{t=0} \right) \right. \right.$$
$$\left. \left. + F\left(h_{t=t_{ch},f} - h_{t=0} \right) \right] + \left(c_s \left(T_{pcm} - T_{t=t_{ch}} \right) \right) \right]$$

(4.65)

$$\sum_{t=t_{ch}}^{t_{st}} \dot{Q}_{gain}\, t = m_{sm} \left[\left(c_f \left(T_{t=t_{st}} - T_{pcm} \right) \right) + \left[(1-F)\left(h_{t_{st},s} - h_{t_{ch}} \right) \right. \right.$$
$$\left. \left. + F\left(h_{t_{st},f} - h_{t_{ch}} \right) \right] + \left(c_s \left(T_{pcm} - T_{t=t_{ch}} \right) \right) \right]$$

(4.66)

$$\sum_{t=t_{st}}^{t_{disch}} \left(\dot{m}_7 h_7 - \dot{m}_8 h_8 \right) t = m_{sm} \left[\left(c_f \left(T_{t=0} - T_{pcm} \right) \right) + \left[(1-F)\left(h_{t=t_{disch},s} - h_{t=t_{st}} \right) \right. \right.$$
$$\left. \left. + F\left(h_{t=t_{disch},f} - h_{t=t_{st}} \right) \right] + \left(c_s \left(T_{pcm} - T_{t=t_{ch}} \right) \right) \right]$$

(4.67)

The overall exergy balance for the ice storage tank can be written as

$$-(\textit{Exergy charged}) + (\textit{Exergy discharged}) + (\textit{Heat loss due to heat losses})$$
$$- (\textit{Exergy destruction}) = (\textit{Exergy accumulation})$$

and

$$-Ex_{ch} + Ex_{disch} + Ex_{Q_{gain}} - Ex_{dest} = \Delta E_{sm}$$

(4.68)

Ex_{ch}, Ex_{disch}, and $Ex_{Q_{gain}}$ can be calculated, respectively, as

$$Ex_{ch} = \int_{t=0}^{t_{ch}} \dot{m}_5(ex_5 - ex_6)dt \quad \text{(J, kJ)}$$

$$Ex_{ch} = \sum_{t=0}^{t_{ch}} \dot{m}_5(ex_5 - ex_6)\, t \quad \text{(J, kJ or kW)}$$

(4.69)

$$Ex_{disch} = \int_{t=0}^{t_{disch}} \dot{m}_7 (ex_7 - ex_8) dt \quad \text{(J, kJ)}$$

$$Ex_{disch} = \sum_{t=0}^{t_{disch}} \dot{m}_8 (ex_7 - ex_8) \, t \quad \text{(J, kJ or kW)}$$

(4.70)

$$Ex_{Q_{gain}} = Q_{gain} \left(1 - \frac{T_0}{T}\right) \tag{4.71}$$

ΔE_{sm} is the exergy change in the PCM and defined as

$$\Delta Ex_{sm} = m_{sm} \left[\left(ex_{initial} - ex_{f@T_{pcm}}\right) + h_{sf} \left(-1 + \frac{T_0}{T_{pcm}}\right) + \left(ex_{s@T_{pcm}} - ex_{final}\right) \right] \tag{4.72}$$

where the first term on the right-hand side of the equation denotes the exergy change due to sensible heat change of the liquid PCM, the second term represents the exergy change due to the phase change, and the last term is the exergy change due to sensible heat change in the solid PCM (if the system goes subcooling after full solidification). The term of ex can be written as

$$ex = (u_i - u_0) - T_0(s_i - s_0)$$

or

$$ex = c(T_i - T_{inital}) - T_0 \ln\left(\frac{T_i}{T_0}\right) \tag{4.73}$$

where "i" denotes the condition of the storage medium in any time.

The overall energy efficiency for the ice storage systems is defined as

$$\eta = \frac{\text{Net recovered energy (cold output)}}{\text{Total charged energy (cold input)}}$$

and

$$\eta_{overall} = \frac{Q_{disch}}{Q_{ch}} = \frac{\sum_{t=t_{st}}^{t_{disch}} \dot{m}_8 (h_7 - h_8)t}{\sum_{t=0}^{t_{ch}} \dot{m}_5 (h_5 - h_6)t} \tag{4.74}$$

Energy efficiencies for operating periods, which are charging, storing, and discharging periods, can be written as

$$\eta_{ch} = \frac{\Delta E_{sm,ch}}{Q_{ch}}$$

$$\eta_{ch} = \frac{m_{sm}\left[\left(c_f\left(T_{t=0} - T_{pcm}\right)\right) + \left[(1-F)\left(h_{t=t_{disch},s} - h_{t=t_{st}}\right) + F\left(h_{t=t_{disch},f} - h_{t=t_{st}}\right)\right] + \left(c_s\left(T_{pcm} - T_{t=t_{ch}}\right)\right)\right]}{\sum_{t=0}^{t_{ch}} \dot{m}_5 (h_5 - h_6)t}$$

(4.75)

$$\eta_{st} = \frac{\Delta E_{sm,st} + Q_{gain}}{\Delta E_{sm,st}}$$

$$\eta_{st} = \frac{m_{sm}\left[\left(c_f\left(T_{t=0} - T_{pcm}\right)\right) + \left[(1-F)\left(h_{t=t_{disch},s} - h_{t=t_{st}}\right) + F\left(h_{t=t_{disch},f} - h_{t=t_{st}}\right)\right] + \left(c_s\left(T_{pcm} - T_{t=t_{ch}}\right)\right)\right] - Q_{loss}}{m_{sm}\left(u_{st@final} - u_{st@initial}\right)}$$

(4.76)

$$\eta_{disch} = \frac{Q_{disch}}{\Delta E_{sm,disch}}$$

$$\eta_{disch} = \frac{\sum\limits_{t=0}^{t_{disch}} \dot{m}_8(h_7 - h_8)t}{m_{sm}\left[\left(c_f\left(T_{t=0} - T_{pcm}\right)\right) + \left[(1-F)\left(h_{t=t_{disch},s} - h_{t=t_{st}}\right) + F\left(h_{t=t_{disch},f} - h_{t=t_{st}}\right)\right] + \left(c_s\left(T_{pcm} - T_{t=t_{ch}}\right)\right)\right]}$$

(4.77)

The overall exergy balance for the ice storage tank can be written as

$$\psi = \frac{Net\ recovered\ cold\ exergy\ capacity}{Total\ charged\ cold\ exergy\ capacity}$$

and

$$\psi_{overall} = \frac{Ex_{disch}}{Ex_{ch}} = \frac{\sum\limits_{t=0}^{t_{disch}} \dot{m}_8(h_7 - h_8)t}{\sum\limits_{t=0}^{t_{ch}} \dot{m}_5(ex_5 - ex_6)t}$$

(4.78)

The exergy efficiencies for charging, storing, and discharging periods can be calculated, respectively, as

$$\psi_{ch} = \frac{\Delta Ex_{sm,ch}}{Ex_{ch}}$$

$$\psi_{ch} = \frac{m_{sm}\left[\left(ex_{initial} - ex_{f@T_{pcm}}\right) + h_{sf}\left(-1 + \frac{T_0}{T_{pcm}}\right) + \left(ex_{s@T_{pcm}} - ex_{final}\right)\right]}{\sum\limits_{t=0}^{t_{ch}} \dot{m}_5(ex_5 - ex_6)t}$$

(4.79)

$$\psi_{st} = \frac{\Delta Ex_{sm,st} + Ex_{Q_{gain}}}{\Delta Ex_{sm,st}}$$

$$\psi_{st} = \frac{m_{sm}\left[\left(ex_{initial} - ex_{f@T_{pcm}}\right) + h_{sf}\left(-1 + \frac{T_0}{T_{pcm}}\right) + \left(ex_{s@T_{pcm}} - ex_{final}\right)\right] - Q_{loss}}{m_{sm}\left(u_{st@final} - u_{st@initial}\right)}$$

(4.80)

$$\psi_{disch} = \frac{Ex_{disch}}{\Delta Ex_{sm,disch}}$$

$$\psi_{disch} = \frac{\sum\limits_{t=0}^{t_{disch}} \dot{m}_8(ex_7 - ex_8)t}{m_{sm}\left[\left(ex_{initial} - ex_{f@T_{pcm}}\right) + h_{sf}\left(-1 + \frac{T_0}{T_{pcm}}\right) + \left(ex_{s@T_{pcm}} - ex_{final}\right)\right]}$$

(4.81)

Illustrative Example 4. Energy and exergy analyses of an encapsulated ice storage tank

Consider an encapsulated ice storage system containing 20,000 capsules, shown in Fig. 4.9. The diameter of the capsule is 11 cm. The volume of the tank is 28 m³, and the diameter and height of the tank are 2.7 and 5 m, respectively. 30% ethylene glycol solution is used as HTF in the system. The system works in a daily cycle, which is 12 h of charging period, 6 h of storing period, and 6 h of discharging period. During the charging period, inlet and outlet temperatures are −10°C and −4°C. At the beginning, the temperature inside the tank is 10°C. At the end of the charging period, it is expected to reduce the temperature of the storage medium to −3°C. Calculate the energy and exergy efficiencies of the overall heat storage process.

Solution: The assumptions to be applied to the problem first defined:

- Reference conditions are $T_0 = 25°C$ and $P_0 = 100$ kPa.
- There is no pressure loss in the tank.
- The temperature inside the tank is uniform.
- The solidification of water/ice occurs at 0°C, and the temperature does not change during the phase change.

The state points and properties of the working fluids are given in Table 4.4.

To calculate the amount of the charged energy, the mass of the storage medium should be determined:

$$m_{sm} = \rho_{ice} V_{cap} N = 916.7 * \frac{4}{3} * \pi * 0.055^3 * 20000 \rightarrow m_{sm} = 12777 \text{ kg}$$

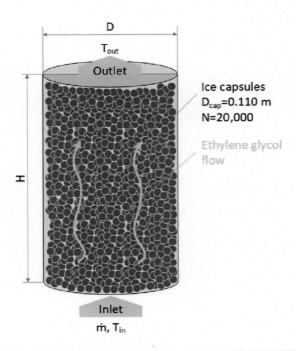

FIG. 4.9

A schematic representation of an encapsulated ice storage tank.

Table 4.4 The State Point Table for the Working Fluids in illustrative example 4.

State Point	Working Fluid	T (°C)	P (kPa)	h (kJ/kg)	ex (kJ/kg)
$T_{in,ch}$	Ethylene glycol	−10	100	−67.88	8.11
$T_{out,ch}$	Ethylene glycol	−4	100	15.34	5.49
$T_{in,disch}$	Ethylene glycol	10	100	−19.27	1.44
$T_{out, disch}$	Ethylene glycol	4	100	−44.38	2.86

The density of the ice is $\rho_{ice} = 916.7$ kg/m^3 at the temperature of 0°C. N is the total number of ice capsules inside the storage tank.

As we know the initial and final states of the storage medium, we can calculate the total charged energy ($Q_{charged}$), also called energy change in the storage medium (ΔE_{sm}).

$$Q_{total,ch} = \Delta E_{sm,ch} = Q_{sen,f,ch} + Q_{pcm,ch} + Q_{sen,s,ch}$$

where $Q_{sen,f,ch}$ is the sensible heat change in water up to the reach melting point. $Q_{pcm,ch}$ is the latent heat change during the solidification. $Q_{sen,s,ch}$ is the sensible heat change in the ice up to the end of charging period. They are calculated as follows:

$$Q_{sen,f,ch} = m_{sm}c_{water}\left(T_{init,ch} - T_{pcm}\right) = 12777 * 4.205 * (10 - 0) \rightarrow Q_{sen,f,ch} = 537,268 \text{ kJ}$$

where c_{water} is the specific heat of the water at the temperature of 5°C, which is the average temperature of the initial and phase change temperatures of the water.

$$Q_{pcm,ch} = m_{sm}h_{sf} = 12777 * 333.6 \rightarrow Q_{pcm,ch} = 4,262,407 \text{ kJ}$$

Here, $h_{sf} = 333.6 \; kJ/kg$ is the phase change enthalpy of the water/ice.

$$Q_{sen,s,ch} = m_{sm}c_{ice}\left(T_{pcm} - T_{final,ch}\right) = 12777 * 2.145 * (0 - (-3)) \rightarrow Q_{sen,s,ch} = 82202 \text{ kJ}$$

Thus, the total charged energy is

$$Q_{total,ch} = 537,268 + 4,262,407 + 82,202 = 4,881,877 \text{ kJ}$$

Charging period: $t_{ch} = 12$ h

MBE for the charging period:

$$\dot{m}_{in,ch} = \dot{m}_{out,ch} = \dot{m}_{ch} \text{ and } m_{sm,init,ch} = m_{sm,final,ch}$$

EBE for the charging period: We can determine the mass flow rate of ethylene glycol from the EBE. Enthalpy values are taken from Table 4.4.

$$\sum_{t=0}^{t_{ch}=12}\left(\dot{m}_{in,ch}h_{in,ch} - \dot{m}_{out,ch}h_{out,ch}\right)t = \Delta E_{sm,ch}$$

$$\dot{m}_{ch} * (-67.88 - 15.34) * 12 * 3600 = 4,881,877$$

$$\dot{m}_{ch} = 1.358 \; (\text{kg}/\text{s})$$

ExBE for the charging period: The exergy destruction in the charging period is determined with ExBE.

$$\sum_{t=0}^{t_{ch}=12} \left(\dot{m}_{in,ch} ex_{in,ch} - \dot{m}_{out,ch} ex_{out,ch} \right) t - Ex_{dest,ch} = \Delta Ex_{sm,ch}$$

where the specific exergies in the parenthesis are calculated as follows, and their values are given in Table 4.4.

$$ex_{in,ch} = c_{ave,ch} \left[(T_{in,ch} - T_0) - T_0 \ln\left(\frac{T_{in,ch}}{T_0}\right) \right]$$

$$ex_{out,ch} = c_{ave,ch} \left[(T_{out,ch} - T_0) - T_0 \ln\left(\frac{T_{out,ch}}{T_0}\right) \right]$$

The exergy changes in the storage medium ($\Delta Ex_{sm,ch}$) can be calculated as

$$\Delta Ex_{sm,ch} = Ex_{f,ch} + Ex_{pcm,ch} + Ex_{s,ch}$$

Here, $Ex_{sm,f,ch}$ is the exergy change due to the sensible heat change in the water, $Ex_{pcm,ch}$ is the exergy change due to the latent heat change during the phase change, and $Ex_{sm,s,ch}$ is the exergy change due to the sensible heat change in the ice.

$$Ex_{f,ch} = m_{sm}(ex_{init} - ex_{pcm,f}) = 12777 * (4.67 - 1.642) = 38,694 \text{ kJ}$$

$$Ex_{pcm,ch} = m_{sm} h_{sf} \left[-1 + \frac{T_0}{T_{pcm}} \right] = 12777 * 333.6 * \left[-1 + \frac{25 + 273.15}{0 + 273.15} \right] = 390,116 \text{ kJ}$$

$$Ex_{s,ch} = m_{sm}(ex_{pcm.s} - ex_{final}) = 12777 * (3.10 - 2.382) = 8020 \text{ kJ}$$

ex_{init} and $ex_{pcm,f}$ are the specific exergies at the beginning of the charging period and solidification, and $ex_{pcm.s}$ and ex_{final} are the specific exergies at the end of the solidification and charging period.

Now, the exergy destruction is calculated from the ExBE.

$$[1.358 * (8.11 - 5.49) * 12 * 3600] + Ex_{dest,ch} = 38,694 + 390,116 + 8020$$

$$Ex_{dest,ch} = 198,009 \text{ kJ}$$

Entropy generation in the charging period is calculated as

$$Ex_{dest,ch} = T_0 S_{gen,ch} \rightarrow S_{gen,ch} = 664.1 \text{ kJ}$$

Storing period: $t_{ch} = 6 \text{ h}$
MBE: There is no mass inlet and outlet in the tank. $m_{sm,init,st} = m_{sm,final,st}$.
EBE for the storing period: Only interaction passing from system boundaries is the heat gain from surrounding to the tank.

$$\sum_{t=0}^{t_{st}} -\dot{Q}_{gain} t = \Delta E_{sm,st}$$

$$3.5 * 6 * 3600 = 4,881,877 - E_{sm,st,final} \rightarrow E_{sm,st,final} = 4,806,277 \text{ kJ}$$

From $E_{sm,st,final}$, the temperature of the storage medium at the end of the storing period can be calculated.

$$E_{sm,st,final} = Q_{sen,f,ch} + Q_{pcm,ch} + m_{sm}c_{ice}\left(T_{final} - T_{st,final}\right)$$

$$4,806,277 = 537,268 + 4,262,407 + 12777 * 2.145 * \left(-4 - T_{st,final}\right) \rightarrow T_{st,final}$$
$$= -0.24°C$$

ExBE for the storing period:

$$-Ex_{Q_{gain}} - Ex_{dest,st} = \Delta Ex_{sm,st}$$

$$Ex_{Q_{gain}} = Q_{gain}\left[1 - \frac{T_{pcm}}{T_0}\right] = 3.5 * 6 * 3600 * \left[1 - \frac{0 + 273.15}{25 + 273.15}\right] = 6339\,\text{kJ}$$

$$\Delta Ex_{sm,st} = Ex_{f,st} + Ex_{pcm,st} + Ex_{s,st}$$

Here, during the storing period, only temperature of the ice increases with the heat gain. Only $Ex_{s,st}$ changes, and $Ex_{f,st}$ and $Ex_{pcm,st}$ are equal to $Ex_{f,ch}$ and $Ex_{pcm,ch}$, respectively. $Ex_{s,st}$ is calculated as

$$Ex_{s,st} = m_{sm}\left(ex_{final,st} - ex_{init,st}\right)$$

$ex_{init,st}$ is equal to ex_{final}.

$$ex_{final,st} = c_p\left[\left(T_{final,st} - T_0\right) - T_0\,ln\left(\frac{T_{final,st}}{T_0}\right)\right] = 2.429\,\text{kJ/kg}$$

$$Ex_{s,st} = 12777 * (3.01 - 2.429) = 607.4\,\text{kJ}$$

From ExBE,

$$Ex_{dest,st} = 1073\,\text{kJ}$$

The entropy generation in the storing period:

$$Ex_{dest,st} = T_0 S_{gen,st} \rightarrow S_{gen,st} = 3.6\,\text{kJ}$$

Discharging period: $t_{ch} = 6$ h
MBE for the discharging period:

$$\dot{m}_{in,disch} = \dot{m}_{out,disch} = \dot{m}_{disch}\ \text{and}\ m_{sm,init,disch} = m_{sm,final,disch}$$

EBE for the discharging period: The remaining energy at the end of storing period is discharged.

$$\sum_{t=0}^{t_{disch}=6}\left(\dot{m}_{in,disch}h_{in,disch} - \dot{m}_{out,ch}h_{out,disch}\right)t = \Delta E_{sm,disch}$$

We can determine the mass flow rate of the HTF with the EBE by getting enthalpy values from Table 4.4.

$$\dot{m}_{disch} * (-19.27 - (-44.38)) * 6 * 3600 = 4,806,277$$

$$\dot{m}_{disch} = 8.863\,\text{kg/s}$$

ExBE for the discharging period:

$$\sum_{t=0}^{t_{disch}=6} \left(\dot{m}_{in,disch} ex_{in,disch} - \dot{m}_{out,disch} ex_{out,disch} \right) t - Ex_{dest,disch} = \Delta Ex_{sm,disch}$$

$$8.863 * (2.86 - 1.44) * 6 * 3600 - Ex_{dest,disch} = 38,694 + 390,116 + 6074$$

$$Ex_{dest,disch} = 157,572 \text{ kJ}$$

The entropy generation in the storing period:

$$Ex_{dest,disch} = T_0 S_{gen,disch} \rightarrow S_{gen,disch} = 528.7 \text{ kJ}$$

The overall energy and exergy efficiencies for the heat storage process:

$$\eta_{overall} = \frac{Total\ discharged\ cold\ capacity}{Total\ charged\ cold\ capacity} = \frac{Q_{total,disch}}{Q_{total,ch}} = \frac{4,806,277}{4,881,877} = 0.98$$

$$\eta_{overall} = \frac{Exegry\ of\ the\ total\ discharged\ cold\ capacity}{Exegry\ of\ the\ total\ charged\ cold\ capacity} = \frac{Ex_{total,disch}}{Ex_{total,ch}} = \frac{153,703}{271,845} = 0.56$$

Illustrative Example 5. Volume fraction change in a latent heat storage unit

Consider a latent heat storage unit filled with 1 kg water, as shown in Fig. 4.10. There is a tube bank flowing an HTF inside. The bulk temperature of the HTF is $-4°C$, and the heat transfer rate from the storage unit to HTF is 0.5 kW. At the beginning of the charging period, the temperature of the storage medium is $0°C$, and its phase is liquid. Calculate the time required to fully solidify. Plot the liquid volume fraction versus time.

Solution:

First, the following assumptions are applied to the problem.

- Reference conditions are $T_0 = 25°C$ and $P_0 = 101.325$ kPa.
- There is no pressure loss in the tube bank.
- The temperature inside the tank is uniform.
- The solidification of water/ice occurs at $0°C$, and the temperature does not change during the phase change.
- The heat flux between HTF and storage medium is uniform.

The temperature of the storage medium is $0°C$. When the HTF flows through the tube bank, water starts to solidify. To determine the charging period, the EBE should be written for the heat storage unit. EBE of the heat storage unit:

$$\dot{m}_{in} h_{in} + \dot{m}_{out} h_{out} = -\Delta E_{sm}$$

Here, $\left(\dot{m}_{in} h_{in} + \dot{m}_{out} h_{out} \right)$ is defined as $\dot{Q}_{flux} = 0.5$ kW. Now, EBE can be written as

$$-\dot{Q}_{flux} t_{ch} = -\Delta E_{sm}$$

$$\Delta E_{sm} = m_{sm} h_{sf} = 1(\text{kg}) * 333.6 \ (\text{kJ}/\text{kg}) \rightarrow \Delta E_{sm} = 333.6 \text{ kJ}$$

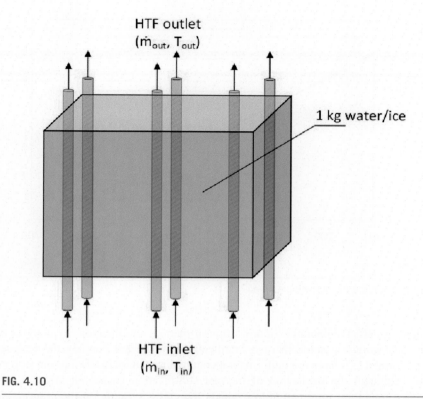

HTF outlet
($\dot{m}_{out}$, T_{out})

1 kg water/ice

HTF inlet
($\dot{m}_{in}$, T_{in})

FIG. 4.10

A schematic representation of the latent heat storage unit consisted 1 kg of water/ice.

$$-\dot{Q}_{flux}\, t_{ch} = -\Delta E_{sm} \rightarrow -0.5\,(kW) * t_{ch}(s) = 333.6\,(kJ) \rightarrow t_{ch} = 667.2\ \text{s or } 11.12\ \text{min}$$

To determine the volume fraction versus time in the heat storage unit, ΔE_{sm} can be written in detail form like in Eq. (4.61):

$$-\dot{Q}_{flux}\, t = -m_{sm}\left[(1-F)(h_s - h_0) + F(h_f - h_0)\right]$$

Here, $h_s = -332.6$ kJ/kg and $h_f = 0.06101$ kJ/kg. $h_0 = h_f$, initial condition of the storage medium is water at the temperature of $0°C$. From the aforementioned equation, time (t) can be determined for each second. Fig. 4.11 demonstrates the liquid volume fraction against time. As the heat transfer rate between the storage medium and HTF is assumed constant, the volume fraction change with time is linear.

4.6 Energy and Exergy Analyses of Thermally Stratified Heat Storage System

There is no uniform temperature distribution in almost all heat storage systems. In other words, there is a temperature gradient in the heat storage unit. This temperature

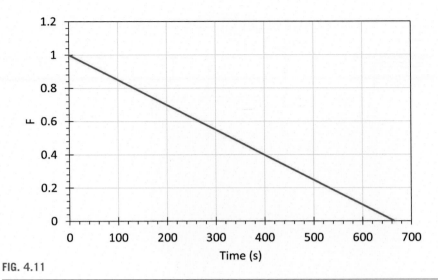

FIG. 4.11

The liquid volume fraction versus time.

gradient is called temperature stratification or thermal stratification. In analyses of many heat storage systems, thermal stratification is neglected. Namely, the analyses are performed as if there is a uniform temperature distribution inside the heat storage units. Although the neglection of thermal stratification causes an error, it still provides important information to evaluate the entire system. On the other hand, it should be included in the calculations to achieve more accurate results, especially when the thermal stratification is dominant.

The degree of thermal stratification is one of the most significant and effective assessment criteria for heat storage units with temperature gradients. Various parameters are defined to determine the degree of thermal stratification, such as MIX number, stratification number, Richardson number, and so on. Also, many methods and devices improving the degree of thermal stratification are introduced in the literature. In such systems, thermal stratification should be taken into consideration with proper evaluation methods. Hot water tanks, cold storage tanks, ice storage tanks, solar ponds, and packed bed heat storage volumes are the common examples of thermally stratified heat storage units.

Energy and exergy analyses are a convenient method to evaluate the thermally stratified heat storage systems. While energy analysis defines the heat and work interactions crossing the system boundaries and their impact on the internal energy of the storage medium, exergy analysis provides the evaluation of the quality of heat storage. Also, it identifies the heat storage systems with different temperatures but the same energy and exergy levels. It provides the capacity of the heat storage according to reference conditions (dead state), too.

Total energy and exergy of heat storage systems can be calculated as

$$E = \int_m (e)\, dm \tag{4.82}$$

$$Ex = \int_m (ex)\, dm \tag{4.83}$$

where "e" and "ex" denote the specific energy and exergy, respectively. They can be written as

$$e = u(T) - u_0 \quad \text{or} \quad e = c(T - T_0) \tag{4.84}$$

$$ex = (u(T) - u_0) - T_0(s(T) - s_0) \quad \text{or} \quad ex = c(T - T_0) - T_0 \ln\left(\frac{T}{T_0}\right) \tag{4.85}$$

Note that kinetic and potential energy changes are neglected in Eqs. (4.84) and (4.85).

In most thermally stratified heat storage units, the temperature gradient in the unit can be assumed one-dimensional. So, the temperature in the heat storage unit only changes with the height (h).

Also, as the cross-sectional area of the unit (A) does not change with the height, the mass of the heat storage unit is defined as

$$dm = \frac{m}{H}\, dh \tag{4.86}$$

Thus, E and Ex can be written as

$$E = \frac{m}{H} \int_0^H e(h)\, dh \tag{4.87}$$

$$Ex = \frac{m}{H} \int_0^H ex(h)\, dh \tag{4.88}$$

Here, "e" and "ex" are the functions of the height, as the temperature only changes with the height. "e" and "ex" can be written as follows as a function of height.

$$e = u(h) - u_0 \quad \text{or} \quad e(h) = c(T(h) - T_o) \tag{4.89}$$

$$ex = (u(h) - u_0) - T_0(s(h) - s_0) \quad \text{or} \quad ex(h) = e(h) - c\, T_o \ln(T(h) - T_o) \tag{4.90}$$

When Eq. (4.89) is written in Eq. (4.87), the energy is defined to be

$$E = m\, c\, (T_m - T_o) \tag{4.91}$$

where T_m is the temperature of the fully mixed storage medium and can be calculated as

$$T_m = \frac{1}{H} \int_0^H T(h)\, dh \tag{4.92}$$

In other words, T_m is the average temperature of the tank. It should be noted here that Eq. (4.91) also defines the energy content of the fully mixed heat storage medium.

Similarly, Eq. (4.90) is written in Eq. (4.88), and the exergy content of the heat storage medium is

$$Ex = E - m\,c\,T_o\,\ln(T_e\,/\,T_o) \tag{4.93}$$

where T_e is the equivalent temperature of the heat storage medium, and it is defined as

$$T_e = \exp\left[\frac{1}{H}\int_0^H \ln T(h)dh\right] \tag{4.94}$$

Generally, in most cases, T_e is not equal to T_m ($T_e \neq T_m$). Since T_m is independent from the degree of the thermal stratification, T_e is dependent on the degree of the thermal stratification. They are equal once the heat storage medium is fully mixed.

4.6.1 Effect of Temperature Distribution on Energy and Exergy of the Thermally Stratified Heat Storage Systems

As emphasized in the previous section, the vertical temperature distribution directly affects the energy and exergy contents of the heat storage medium. Therefore, it is a significant issue to determine the temperature distribution inside the heat storage medium. In this section, various temperature gradient models to be able to be used for analyzing the thermal stratified heat storage are introduced.

4.6.1.1 Linear temperature gradient model

In the linear temperature gradient model, it is assumed that the temperature inside the tank changes linearly from the tank bottom to the top. It is a quite useful assumption when we only have the inlet and outlet temperatures of the heat storage unit. In such a case, the temperature inside the tank changes linearly from inlet temperature (bottom) to outlet temperature (top). The temperature change is illustrated in Fig. 4.12. Temperature change can be written as in Eq. (4.95).

$$T(h) = \frac{T_{top} - T_{bottom}}{H}h + T_{bottom} \tag{4.95}$$

T_m and T_e can be calculated as follows, respectively:

$$T_m = \frac{T_{top} + T_{bottom}}{2} \tag{4.96}$$

$$T_e = \exp\left[\frac{T_{top}\,\ln(T_{top} - 1) - T_{bottom}\ln(T_{bottom} - 1)}{T_{top} - T_{bottom}}\right] \tag{4.97}$$

Here, T_{bottom} can be T_{inlet}, and T_{top} can be T_{outlet}.

4.6.1.2 Stepped temperature gradient model

In the stepped temperature gradient model, it is assumed that there is an "n" number of layers inside the tank with a constant temperature change. This assumption is useful when temperature measurements are to be gathered from different heights inside the tank. The stepped temperature gradient model is shown in Fig. 4.13. In Fig. 4.13,

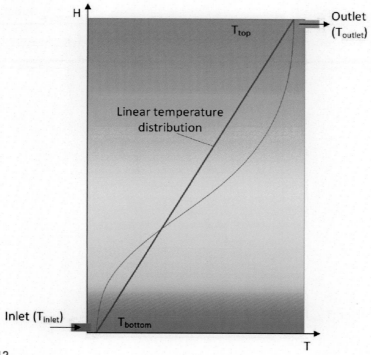

FIG. 4.12

Linear temperature gradient.

while green line shows two-layer stepped temperature gradient model, black line shows "k"-layer stepped temperature gradient model. Temperature change can be defined as the follows:

$$T(h) = \begin{cases} T_1 & h_0 \le h \le h_1 \\ T_2 & h_1 < h \le h_2 \\ \dots \\ T_k & h_{k-1} < h \le h_k \end{cases} \tag{4.98}$$

T_m and T_e can be calculated as the follows, respectively:

$$T_m = \sum_{i=1}^{k} x_j (T_m)_i \tag{4.99}$$

$$T_e = exp\left[\sum_{i=1}^{k} x_i \ln (T_e)_i\right] = \prod_{i=1}^{k} (T_e)_i^{x_i} \tag{4.100}$$

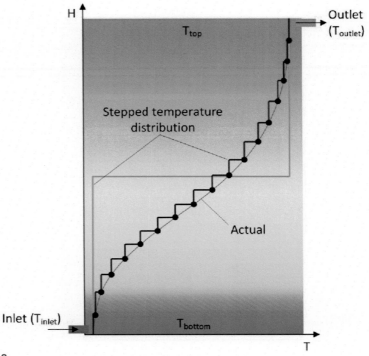

FIG. 4.13

Linear temperature gradient.

Here, the thickness of the layers is identical. m_i is the mass in a layer, and T_i is the temperature of the layer.

4.6.1.3 Continuous-linear temperature gradient model

In this model, as in the stepped temperature gradient model, the heat storage unit is divided into "k" layers, which have same height. The temperature changes linearly in each layer. The continuous-linear temperature gradient model is given in Fig. 4.14. This model can be used when it is possible to obtain temperature measurement from the storage medium in different height, like in stepped temperature gradient model. However, continuous-linear model can provide more precious results than stepped model.

$$T(h) = \begin{cases} T_{1\rightarrow2}(h) & h_0 \leq h \leq h_1 \\ T_{2\rightarrow4}(h) & h_1 < h \leq h_2 \\ \quad\cdots \\ T_{(k-1)\rightarrow k} & h_{k-1} < h \leq h_k \end{cases} \qquad (4.101)$$

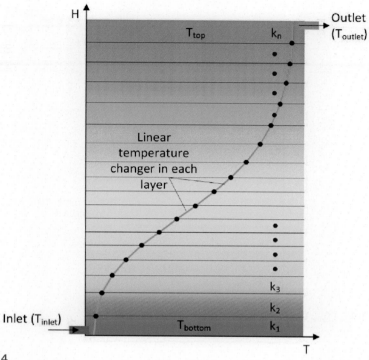

FIG. 4.14

Linear temperature gradient.

Temperature change in a layer can be written as becoming "i" is the bottom of the layer and "$i+1$" is the top of the layer.

$$T_{i \to (i+1)} = \frac{T_{i+1} - T_i}{H_i} h + \frac{h_{i+1} T_i - h_i T_{i+1}}{H_i} \qquad (4.102)$$

where H_i is the height of the layer.

T_m and T_e can be calculated as the follows, respectively:

$$T_m = \sum_{i=1}^{k} x_i (T_m)_i \,, here \ (T_m)_i = \frac{T_i + T_{i+1}}{2} \qquad (4.103)$$

$$T_e = \prod_{i=1}^{k} (T_e)_i^{x_i} \qquad (4.104)$$

Here, $(T_e)_i^{x_i}$ is the equivalent temperature of a layer, and it can be calculated as in Eq. (4.97).

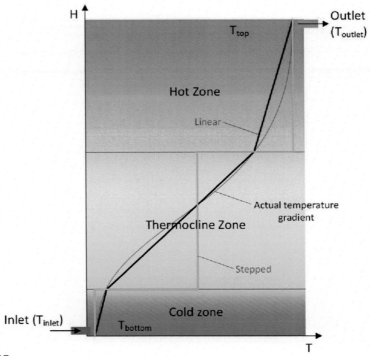

FIG. 4.15

Three-zone temperature gradient.

4.6.1.4 Three-zone temperature gradient model

This temperature gradient model is quite similar to the stepped and continuous-linear temperature gradient models. In this model, the storage medium is divided into three zones. Dividing heat storage units into three zones, especially in sensible heat storage units, is the common technique to evaluate thermal stratification. These zones are cold, thermocline, and hot zones. The degree of thermal stratification is defined over the thickness of these zones. Fig. 4.15 illustrates the three-zone temperature gradient model. The temperature change in the zones can be stepped or linear. This model is useful when the thermal stratification is defined with three zones. It can provide useful information regarding which zone has more energy and exergy contents.

Temperature equations for the three-zone temperature gradient model are the same as introduced previously. When the temperature change in the zones is assumed constant, Eqs. (4.98)−(4.100) are valid for analysis. When the temperature change in the zones changes linearly, Eqs. (4.101)−(4.104) are used in the analysis.

A comprehensive example related to thermal stratification and its improvement is presented in the case studies. Also, temperature gradient models are compared to determine the energy and exergy contents and the degree of thermal stratification.

4.7 Heat Transfer Analyses

Heat transfer analyses are an essential part of analyzing heat storage systems. First and foremost, all heat storage systems have interactions with their surroundings, which are heat loss or heat gain. To determine the heat transfer rates between the heat storage medium and its surrounding, a proper heat transfer analysis should be performed. Besides, an HTF is used in almost all heat storage systems, and heat transfer occurs between the storage medium and HTF. Again, heat transfer analysis is the only tool for determining the heat transfer rate.

In the previous section, we have discussed the thermodynamic analysis methods with illustrative examples. Thermodynamic is interested in the initial/final or inlet/outlet states of the processes and systems. Thermodynamic analyses provide to evaluate the system performance based on data obtained from experimental, numerical, or analytic works. Heat transfer analyses are an indispensable part of these studies to produce data to be used in thermodynamic analyses and to analyze and evaluate the system elements.

Thermal processes by which heat flux occurs from one point to another are seen in all energy systems and applications, such as heating, cooling, evaporation, and condensation of the materials. To analyze these processes and provide necessary data for performance analysis such as thermodynamic analysis or design the systems and optimize processes, physical phenomena and practical aspects of the heat transfer should be applied to the problems accordingly.

Heat transfers occur in three modes, as illustrated in Fig. 4.16:

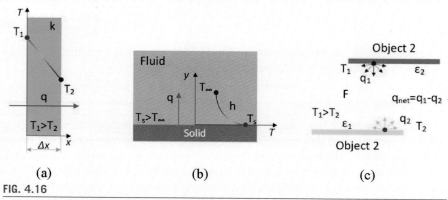

FIG. 4.16

Representation of heat transfer modes: (a) conduction, (b) convection, and (c) radiation.

- Heat conduction
- Heat convection
- Heat radiation

4.7.1 Heat Conduction

Heat conduction, also named thermal conduction, is the heat flux in a body due to temperature difference and occurs by the movement of electrons within a body. Conduction heat transfer continues until the body reaches the thermal balance conditions. The conduction heat transfer also occurs when the solid bodies in difference temperature contact each other. The heat conducts via mobile electrons. When the number of mobile electrons increases, the heat conduction capability of the body is higher; in other words, there is a correlation between heat conductivity and the number of mobile electrons. Therefore, metals conduct heat better than plastics or wood. When the heat conductivity mechanism is compared with the electrical conductivity, it is seen a powerful correlation between heat conductivity and electrical conductivity.

Heat conduction is defined with Fourier's law of heat conduction, which states as the rate of heat flow is proportional with the cross-sectional area and the temperature difference in the homogenous solid body a direct proportional. Fourier's law of heat conduction is written as follows:

$$q_x'' = -k\frac{dT}{dx} \left(\frac{W}{m^2}\right) \quad \text{or} \quad Q_x = -kA\frac{dT}{dx} \tag{4.105}$$

Here, "k" is the coefficient of heat conductivity or thermal conductivity or heat conductivity. The unit of heat conductivity is W/m^2 K or W/m^2°C. All materials have a heat conductivity, which changes with temperature. However, it is assumed as constant in many practical applications. q_x'' is the heat transfer rate in the x-direction per unit area normal to heat flow direction. Under the steady-state conditions, shown in Fig. 4.16, the temperature gradient can be defined as

$$\frac{dT}{dx} = \frac{T_1 - T_2}{L} \quad \rightarrow \quad q_x'' = -k\frac{T_1 - T_2}{\Delta x} \quad \text{or} \quad q_x'' = -k\frac{\Delta T}{\Delta x} \tag{4.106}$$

Note that the last two equations in Eq. (4.106) are called heat flux, which also defines the rate of heat transfer per area.

4.7.2 Heat Convection

When a solid surface has contact with a fluid at a different temperature, this type of heat transfer is called the convection heat transfer. There are two modes of convective heat transfer, which are natural convection and forced convection, defined up to the type of flow. In the forced convection, the fluid zone is forced to flow with a mechanical device such as a pump and fan, or natural flows such as atmospheric wind or river flows. In natural convection, the flow occurs naturally due to buoyancy forces in the fluid. As an example, consider a hot plate contact with cold fluid, as shown in

Fig. 4.16b. When fluid molecules hit the plate surface, the density of molecules changes and that change causes the flow due to the buoyancy forces. For the same fluid and plate at the same temperature conditions, the heat transfer rates are higher in forced convection than the natural convection.

Newton's law of cooling states that the heat transfer from a solid surface to a fluid is proportional to the difference between the surface and fluid temperatures. Convective heat transfer is stated as

$$q = h(T_s - T_\infty)\left(\frac{W}{m^2}\right) \quad \text{or} \quad Q = hA(T_s - T_\infty)(W) \tag{4.107}$$

where "h" denotes the convection heat transfer coefficient. T_s and T_∞ are the temperatures of surface and fluid, respectively. A is the total heat transfer surface area.

Dimensionless numbers are used in convective heat transfer analyses. Some of the most significant heat transfer dimensionless parameters are listed in Table 4.5. Convective heat transfer is more complicated than conductive heat transfer as both solid and fluid are in contact with each other. Many correlations and empirical equations are introduced in heat transfer books. For more information regarding dimensionless numbers, correlations, and empirical equations, please take a glance at heat transfer books.

4.7.3 Heat Radiation

All objects emit radiant energy in all directions unless their temperature is absolute zero. In radiation heat transfer, the heat transfers via in the form of electromagnetic waves or by photons. Unlike conduction and convective heat transfers, the heat flows

Table 4.5 Various Dimensionless Numbers Used in Heat Transfer Analysis.

Dimensionless Number	Equation	Application
Biot number (Bi)	$Bi = \frac{hL}{k_s}$	Steady- and unsteady-state conduction
Fourier number (Fo)	$Fo = \frac{\alpha t}{L^2}$	Unsteady-state conduction
Grashof number (Gr)	$Gr = \frac{g\beta(T_s - T_\infty)L^3}{\nu^2}$	Natural convection
Rayleigh number (Ra)	$Ra = Gr * Pr$	Natural convection
Nusselt number (Nu)	$Nu = \frac{hL}{k_f}$ or $\frac{hD}{k_f}$	Natural, forced, boiling, or condensation heat transfer
Peclet number (Pe)	$Pe = \frac{VL}{\alpha} = Re_L Pr$	Forced convection in lower Pr number
Prandtl number (Pr)	$Pr = \frac{c_p \mu}{k} = \frac{\nu}{\alpha}$	Natural, forced, boiling or condensation heat transfer
Reynolds number (Re)	$Re = \frac{VL}{\nu}$ or $\frac{VD}{\nu}$	Forced convection
Stanton number (St)	$St = \frac{h}{\rho V c_p} = \frac{Nu}{Re\,Pr}$	Modified Nusselt number

Modified from Ref. [2].

both from hot to cold and from cold to hot surfaces. Therefore, the amount of heat transfer in the radiation is called net radiation heat transfer. The direction of the net radiation heat transfer is from the hot surface to the cold surface. Therefore, the amount of heat transfer in the radiation is called net radiation heat transfer. The direction of the net radiation heat transfer is from the hot surface to the cold surface. In the radiation heat transfer, the properties of the surface, which are reflectivity (ρ), absorptivity (α), and transmissivity (τ), have a great impact on the heat transfer rate. There is the following relation between the properties of the surface:

$$\rho + \alpha + \tau = 1 \tag{4.108}$$

If there is no transmission in the body, in other words, if the body is opaque ($\tau = 0$),

$$\rho + \alpha = 1 \tag{4.109}$$

Radiation heat transfer can be defined as follows:

$$q_{12} = \varepsilon_1 \sigma F_{12}\left(T_1^4 - T_2^4\right)\left(W/m^2\right) \quad \text{or} \quad Q_{12} = \varepsilon_1 \sigma A_1 F_{12}\left(T_1^4 - T_2^4\right)(W) \tag{4.110}$$

Here, ε_1 is the emissivity of the surface. F_{12} is the view factor which is specific for the geometrical conditions of the surfaces. σ is the Stephan–Boltzmann constant which has a value of $\sigma = 5.67 \times 10^{-8}$ (W/m^2 K^4).

Illustrative Example 6. Determination of heat transfer rate in a cold heat storage tank

Consider an encapsulated ice storage tank as shown in Fig. 4.17. HTF flows in the tank, and there is the still air outside of the tank. The tank is insulated with a rubber insulation layer. The flow rate, inlet, and outlet temperatures of HTF are given in Table 4.6. Calculate heat transfer rates for charging and discharging periods.

Solution:

Forced heat convection occurs inside the tank as there is the HTF flow, and the natural heat convection takes place in the outside of the tank. Heat conduction occurs in the tank wall and insulation. A common approach to finding the heat gain or loss from a heat storage tank is to first calculate the thermal resistance, which is defined as follows:

$$R_T = \frac{1}{h_{in}} + \frac{\ln\left(r_2/r_1\right)}{2\pi k_{tank}} + \frac{\ln\left(r_3/r_2\right)}{2\pi k_{insulation}} + \frac{1}{h_{out}}$$

Here, the most critical issue is to calculate the values of h_{in} and h_{out}. Heat convection coefficient (h) is calculated as

$$h = \frac{Nu\ k_f}{D}$$

where Nu is the Nusselt number, k_f is the heat conduction coefficient of the fluid at the working conditions, and D is the diameter of the volume studied. For plane walls length (L) is used instead of diameter (D).

Nusselt number (Nu) is the ratio of convective heat transfer to conductive heat transfer across a boundary. It is used for defining the heat transfer rate. In the

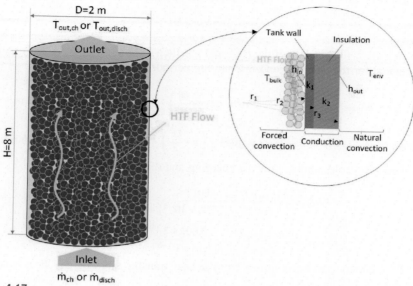

FIG. 4.17

A schematic representation of the heat transfer mechanism in an ice storage tank.

Table 4.6 The Working Temperatures and Thermophysical Features of the Heat Transfer Fluid.

Stream	Temperature (°C)	Bulk Temperature (°C)	Mass Flow Rate (kg/s)	ρ (kg/m³)	ν (m²/s)
$T_{in,ch}$	−6	−4.5	37.78	1046	4.92×10^{-6}
$T_{out,ch}$	−3				
$T_{in,disch}$	10	7	28.97	1043	3.18×10^{-6}
$T_{out,disch}$	4				

literature, there are many correlations of Nu number defining the different problems various conditions. For a heat transfer problem, Nu number equation should be selected properly.

For this problem, Nusselt number should be calculated separately for inside and outside of the tank as the type of heat convection is different. *Nu* number for inner surface of the tank can be calculated as

$$Nu_{in} = \frac{h_{in}D}{k_{HTF}} = 0.023 \, Re^{0.4} \, Pr^{0.4}$$

Reynold number for *HTF* flow inside the tank is calculated as

$$Re = \frac{V_{HTF}\, D}{\nu}$$

where V_{HTF} is the velocity of the *HTF* and ν is the viscosity of the *HTF*. V_{HTF} can be calculated for the charging and discharging periods, respectively, as

$$\dot{m}_{ch} = \rho A_{tank} V_{HTF,ch} \rightarrow 37.78 = 1046 * \left(\pi * \frac{D^2}{4} \right) * V_{HTF,ch} \rightarrow V_{HTF,ch} = 0.012 \text{ m/s}$$

$$\dot{m}_{disch} = \rho A_{tank} V_{HTF,disch} \rightarrow 28.97 = 1043 * \left(\pi * \frac{D^2}{4} \right) * V_{HTF,ch} \rightarrow V_{HTF,disch} = 0.009 \text{m/s}$$

Thus, Reynold numbers for the charging and discharging periods:

$$Re_{ch} = \frac{V_{HTF,ch}\, D}{\nu_{ch}} = \frac{0.012 * 2}{(4.92 * 10^{-6})} \rightarrow Re_{ch} = 4673$$

$$Re_{disch} = \frac{V_{HTF,disch}\, D}{\nu_{disch}} = \frac{0.009 * 2}{(3.18 * 10^{-6})} \rightarrow Re_{disch} = 5568$$

Pr numbers can be obtained from a heat transfer table or any heat transfer database.

$$Pr_{ch} = 42.48$$

$$Pr_{disch} = 26.92$$

Now, as *Re* and *Pr* numbers are obtained, Nu numbers can be calculated as

$$Nu_{in.ch} = 0.023\, 4673^{0.4}\, 42.48^{0.4} \rightarrow Nu_{in.ch} = 88.85$$

$$Nu_{in.disch} = 0.023\, 5568^{0.4}\, 26.92^{0.4} \rightarrow Nu_{in.ch} = 85.18$$

Then,

$$Nu_{in.ch} = \frac{h_{in.ch} D}{k_{HTF,ch}} \rightarrow 88.85 = \frac{h_{in.ch} * 2}{0.4416} \rightarrow h_{in.ch} = 19.62 \left(W / m^2 K \right)$$

$$Nu_{in.disch} = \frac{h_{in.disch} D}{k_{HTF,disch}} \rightarrow 88.85 = \frac{h_{in.disch} * 2}{0.4416} \rightarrow h_{in.disch} = 19.28 \left(W / m^2 K \right)$$

As seen from these results, there is no significant change in heat convection coefficient for the charging and discharging periods, although the temperature of the HTF changes substantially. The average value of two heat convection coefficients can be used for determining the thermal resistance.

$$h_{in} = \frac{\left(h_{in,ch} + h_{in,disch} \right)}{2} = \frac{19.62 + 19.28}{2} \rightarrow h_{in} = 19.45 \left(W / m^2 K \right)$$

For the outside of the tank, *Nu* number can be written as

$$Nu_{out} = \frac{h_{out} D}{k_{air}} = 0.59 * \left(Gr * Pr \right)^{0.25}$$

Grashof (Gr) number can be calculated as in Table 4.5 and found to be 1.492×10^{11}. Pr number is $Pr = 0.7293$.

$$Nu_{out} = 0.59 * \left(1.492 * 10^{11} * 0.7293\right)^{0.25} \rightarrow Nu_{out} = 313.2$$

$$Nu_{out} = \frac{h_{out}D}{k_{air}} \rightarrow 313.2 = \frac{h_{out} * 2}{0.025} \rightarrow h_{out} = 3.936\left(W / m^2K\right)$$

Thermal resistance from outside to inside of the tank is finally calculated as

$$\frac{1}{R_T} = \frac{1}{h_{in}} + \frac{\ln(r_2/r_1)}{2\pi k_{tank}} + \frac{\ln(r_3/r_2)}{2\pi k_{insulation}} + \frac{1}{h_{out}}$$

Here, k_{tank} and $k_{insulation}$ are the heat conductivity of the tank wall and insulation material, which are $k_{tank} = 21.75$ kW/mK and $k_{insulation} = 0.035$ kW/mK. Heat conduction given and convection calculated are written into the previous equation.

$$\frac{1}{R_T} = \frac{1}{19.45} + \frac{ln(r_2/r_1)}{2 * \pi * 21.75} + \frac{ln(r_3/r_2)}{2 * \pi * 0.035} + \frac{1}{3.936}$$

$$R_T = 1.9\left(kW / m^2K\right)$$

This thermal resistance values for all working conditions, which are charging, storing, and discharging, are approximately the same. Now, we can calculate the heat transfer rate from the following equation:

$$\dot{Q}_{gain,ch} = A_{tank}R_T\left(T_\infty - T_{bulk,ch}\right) = 56.55 * 1.9 * (20 - (-4.5)) \rightarrow \dot{Q}_{gain,ch} = 2632 \text{ kW}$$

$$\dot{Q}_{gain,disch} = A_{tank}R_T\left(T_\infty - T_{bulk,disch}\right) = 56.55 * 1.9 * (20 - 7) \rightarrow \dot{Q}_{gain,disch} = 1397 \text{ kW}$$

where A_{tank} is the surface area of the tank, T_∞ is the environmental temperature, and T_{bulk} is the bulk temperature of the HTF at the working condition. When the working condition is considered, only T_{bulk} changes with the periods.

Discussion: In illustrative examples 4 and 5, we have not considered heat loss and gain between control volume and environment. As observed from this illustrative example, it should be included in analyses by considering each operating condition. Heat transfer analysis of the heat storage tanks is not limited to the present example. As mentioned before, there are many correlations for different conditions and working fluids. Proper equations, correlation, and assumption should be applied to the analyses.

Illustrative Example 7. Determination of heat transfer rate in a tube bank

Consider a tube bank filled with water as the storage unit, which is used in the solar heating system. Total volume of storage medium is 1500 L. Hot air that is heated in the solar collector flows through the tubes, and energy is charged into the water, as demonstrated in Fig. 4.18a. Calculate the amount of the energy charged in the heat storage unit considering in-line and staggered arrangements of the tube bank. Parameters to be studied are given in Table 4.7.

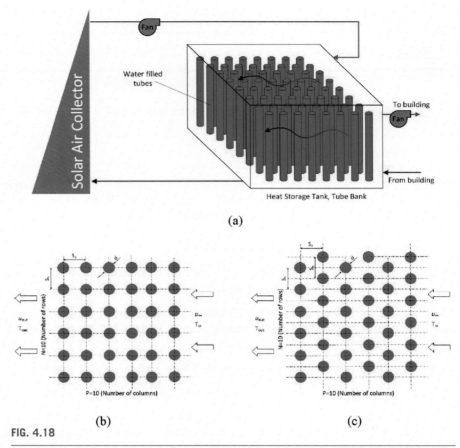

FIG. 4.18

The schematic view of the solar heating system integrated with sensible heat storage unit consisting of tube bank: (a) the system diagram of the system, (b) in-line, and (c) staggered arrangements of the tube bank.

Modified from Ref. [3].

Table 4.7 Parameter Studied in the illustrative example 7.

Parameter	Value
Tube diameter, d, (m)	60, 80, 150 mm
Arrangement	In-line and staggered
S_N	1.25d, 1.50d, and 2.00d
S_P	1.25d, 1.50d, and 2.00d
Inlet velocity, u_∞, (m/s)	2, 4, 6, 8, 10
Inlet temperature, T_{in}, (°C)	50, 70, 90

Adapted from Ref. [3].

Solution:

There is a forced convection between air and tube banks; therefore, the heat transfer coefficient between air and tube bank should be determined. The arrangement of the tube bank is shown in Fig. 4.18b and c. For the tube banks, Nusselt number can be calculated as [4,5]

$$Nu_{(d,f)} = \frac{h\,d}{k_f} = C\,Re_{d,max}^{n}\,Pr_f^{1/3}$$

where C and n are the constant coefficient to be taken from heat transfer table in Ref. [4]. To calculate $Re_{d,max}$, u_{max} should be calculated first. u_{max} for the in-line and staggered arrangements are calculated, respectively, as

$$u_{max,in-line} = u_{\infty}\left(S_n / (S_n - d)\right)$$

$$u_{max,staggered} = \frac{u_{\infty}(S_n/2)}{\left((S_n/2)^2 + S_p^2\right)^{1/2} - d}$$

After calculating u_{max}, $Re_{d,max}$ can be calculated as

$$Re_{d,max} = \frac{\rho_f\,u_{max}\,d}{\mu_f}$$

After calculating $Re_{d,max}$, the Nusselt number is calculated by taking Pr from a heat transfer table or calculating with the equation in Table 4.5. When the Nusselt number is obtained, the heat convection coefficient can easily be calculated. The charged energy to the storage unit is calculated as

$$\dot{Q}_{charged} = \dot{m}c_{p,air}(T_{in} - T_{out})$$

where $c_{p.air}$ is the specific heat of the air. T_{in} and T_{out} are the inlet and outlet temperatures in the storage unit, respectively.

The stored energy in the storage medium is also calculated as

$$\dot{Q}_{stored} = hA_{total}(T_{sur} - T_{bulk})$$

Here, T_{sur} is the surface temperature of the tubes and T_{bulk} is the bulk temperature of the air, which is $T_{bulk} = (T_{in} + T_{out})/2$. When the heat loss from the heat storage unit is neglected, the charged energy will be equal to the stored energy.

$$\dot{Q}_{charged} = \dot{Q}_{stored}$$

T_{out} can be calculated with the previous equation. The results obtained from the heat transfer analysis are given in Tables 4.8 and 4.9, respectively.

Discussion: As seen from the results, there is solid relation between the heat transfer coefficient and the design of the tube bank. To maximize the stored energy in a certain storage volume or minimize the storage volume in a certain amount of energy storage, heat transfer analysis should be applied to the problem. Note that, in this problem, the heat loss is neglected to reduce the complexity of the analysis. For more information and results regarding the aforementioned problem, please see Ref. [3].

Table 4.8 Results for In-Line Arrangement $S_N = 2.0$ and $S_P = 2.0$

u_∞ (m/s)	2	4	6	8	10
$T_{in} = 50°C$					
$Re_{d,max}$	11,985.8	23,971.6	35,957.5	47,943.3	59,929.1
$Nu_{d,f}$	85.7	132.8	171.6	205.8	237.0
H (W/m²K)	39.4	61.1	78.9	94.7	109.0
$\dot{Q}_{stored}$ (kW)	127,681.0	202,000.7	263,627.4	318,191.4	368,025.5
$T_{in} = 70°C$					
$Re_{d,max}$	10,883.2	21,766.3	32,649.5	43,532.6	54,415.8
$Nu_{d,f}$	80.6	124.9	161.4	193.6	223.0
H (W/m²K)	39.2	60.8	78.6	94.2	108.5
$\dot{Q}_{stored}$ (kW)	198,690.9	314,679.9	410,899.4	496,110.6	573,946.7
$T_{in} = 90°C$					
$Re_{d,max}$	9920.3	19,840.7	29,761.0	39,681.4	49,601.7
$Nu_{d,f}$	76.0	117.8	152.3	182.6	210.3
H (W/m²K)	39.0	60.5	78.2	93.7	107.9
$\dot{Q}_{stored}$ (kW)	268,183.0	425,184.0	555,478.9	670,892.6	776,332.6

Modified from Ref. [3].

Table 4.9 Results for In-Line Arrangement $S_N = 1.25$ and $S_P = 1.25$

u_∞ (m/s)	2	4	6	8	10
$T_{in} = 50°C$					
$Re_{d,max}$	8555.0	17,110.1	25,665.1	34,220.1	42,775.1
$Nu_{d,f}$	78.8	115.8	145.1	170.3	192.8
H (W/m²K)	38.3	56.4	70.6	82.9	93.8
$\dot{Q}_{stored}$ (kW)	188,945.9	286,880.3	364,885.7	432,168.5	492,433.2
$T_{in} = 70°C$					
$Re_{d,max}$	8163.6	16,327.1	24,490.7	32,654.3	40,817.9
$Nu_{d,f}$	76.8	112.8	141.4	165.9	187.8
H (W/m²K)	38.4	56.4	70.7	82.9	93.9
$\dot{Q}_{stored}$ (kW)	222,714.8	338,445.9	430,651.8	510,193.8	581,445.1
$T_{in} = 90°C$					
$Re_{d,max}$	7798.2	15,596.4	23,394.5	31,192.7	38,990.9
$Nu_{d,f}$	74.8	110.0	137.8	161.7	183.1
H (W/m²K)	38.4	56.5	70.7	83.0	94.0
$\dot{Q}_{stored}$ (kW)	256,278.5	389,785.7	496,183.4	587,980.5	670,216.6

Modified from Ref. [3].

In the heat transfer analysis section, we have presented only two illustrative examples to emphasize the significance of heat transfer analyses in heat storage systems. The methodology for all heat transfer analyses is similar to the methodology presented in the examples. However, specific equations, correlations, and assumptions should be applied to the specific problem. The main objective of the illustrative examples regarding heat transfer analysis is to give the fundamental aspect to solve the heat transfer problems and their importance of the performance of heat storage systems.

4.7.4 Heat Transfer Analysis in Porous Medium

Heat transfer analysis in porous mediums is a critical issue for the heat storage systems as many of them can be modeled as porous mediums. A porous medium is a material containing pores, as shown in Fig. 4.19. In other words, there are gaps in a solid material, or there are solid particles in a fluid material. Namely, the pores

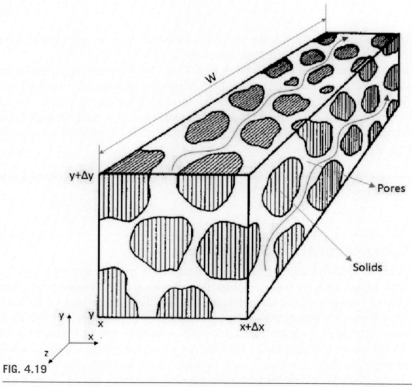

FIG. 4.19

The schematic view of porous medium and averaging of the pore velocity distribution.

Modified from Ref. [6].

are filled with fluid. When the pores are connected to each other, it is called the open-cell porous material. Otherwise, when the pores are not connected to each other, it is called the closed-cell porous material. A porous medium is most characterized by porosity (φ) and permeability (K). Porosity (φ) is defined as the ratio of the void volume to total volume. Permeability (K) quantifies the ability of the formation to transmit fluids. Porosity (φ) is formulated as the follows:

$$\varphi = \frac{V_{void}}{V_{total}} \tag{4.111}$$

Many heat storage systems can be modeled as the porous medium, as they consist of comparatively lower volume solid volume than the total volume of heat storage, such as rock beds and encapsulated heat storage tanks. The storage medium is the solid particles like in rock beds or is kept in solid capsules like in encapsulated heat storage tanks. Therefore, heat transfer analysis of porous mediums is a critical issue to optimize the performance of the system.

The first attempt for the investigation in a fluid flow in a porous medium has been done by Henry Darcy, called Darcy law.

$$u = \frac{K}{\mu}\left(-\frac{dP}{dx}\right) \tag{4.112}$$

where K is the permeability. μ is the viscosity of the fluid flows in pores and P is the pressure, and x is the distance in flow direction.

Many heat storage systems can be modeled with heat transfer equations in the porous medium. 2-D heat transfer equation in cylindrical coordinates for porous medium (as shown in Fig. 4.20) is defined in Eq. (4.113). When Eq. (4.113) is solved, the temperature gradient inside the tank is obtained. Thus, it is possible to perform thermodynamic analysis of the porous medium. A detailed illustrative example is presented in the case studies.

$$\varphi \rho C u \frac{\partial T}{\partial x} = \frac{1}{r}\frac{\partial}{\partial x}\left(kr\frac{\partial T}{\partial r}\right) + \frac{\partial}{\partial x}\left(k\frac{\partial T}{\partial x}\right) + HA_{bed}\left(T_{pcm} - T\right) \tag{4.113}$$

4.8 Computational Fluid Dynamics

Computational fluid dynamics (CFD) is a critical tool to model a system or process physically. Today, there are many commercial and open-source CFD software. With CFD software, it is possible to solve not only one specific problem but also different physical problems simultaneously. When it comes to CFD programs, the first thing that comes to mind is programs that solve fluid mechanics and heat transfer problems. Today, many CFD programs have turned to multiphysics programs, which offer to combine fluid mechanics, heat transfer, chemical processes, solid mechanics, etc. Namely, many hard-to-solve physical problems can be solved with such software at the same time. Also, the effect of one physical problem on another

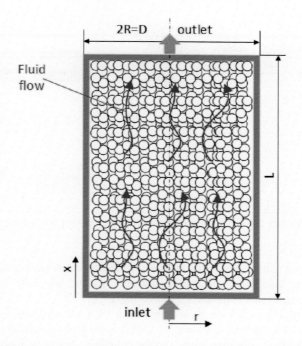

FIG. 4.20

The illustration of the fluid flow in the porous medium in 2-D cylindrical coordinates.

problem can be investigated numerically. For instance, while it is possible to study fluid flow and heat transfer characteristics in a heat exchanger with a CFD analysis, the effect of temperature gradient obtained by CFD on the stress in the solid part of the heat exchanger can be observed. Another example, the chemical process in battery cells or packs can be modeled, and the charging and discharging characteristics of the battery cells or packs can be investigated. Simultaneously, the effect of the chemical process on the temperature gradient in the battery can be examined. Moreover, the cooling processes of the battery and their effects on the charging and discharging characteristics can be modeled simultaneously. Consequently, CFD and multiphysics software are essential tools for engineering, R&D, and technology development processes. Here, some applications of CFD in heat storage systems are introduced to the readers.

Differential equation sets that are specific for the problem are solved by discretizing the differential equations into the algebraic sets of equations to evaluate processes according to the initial and boundary conditions set at the beginning of the analysis. Governing equations are specific to the problem and its coordinate system. For example, if a turbulence model is required for the analysis, the set of equations specific for the selected model is included in the analysis. If the phase changing is to be analyzed, a set of equations used in the phase modeling is added to analyses to

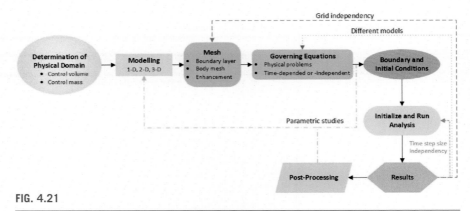

FIG. 4.21

The flow diagram for the CFD analysis procedure. *CFD*, computational fluid dynamics.

solve. Analysis can be performed time-dependent or time-independent. To address the governing equations, the user manuals, tutorials, and help modules of the CFD and multiphysics software should be used.

Fig. 4.21 demonstrates the flow diagram for the CFD analyses. In the numerical analyses, first, the physical domain (also called system) to be studied should be determined. The body staying inside the system boundary should be modeled with solid modeling software or embedded modules in CFD software. The control volume can be 1-D, 2-D, and 3-D. To reduce the complexity of the geometry, symmetry or axisymmetric conditions can be applied. After geometrical modeling, the body is divided into elements (also called mesh) that connected each other. The structure of the elements connected to each other is named the grid structure. One of the critical parameters for the accuracy of the results is the number of elements. The number of elements directly depends on the element size. To achieve more accurate results, the smaller size element should be used. However, the smaller grid size does not always provide better results. In other words, more than a certain number of elements do not affect the results very much. Therefore, the grid independency process should be added to the analysis procedure. An illustration of the grid independency process is shown in Fig. 4.22. In the grid independency process, the number of elements is increased (the size of elements reduced) step by step, and the results are watched. When the results do not change significantly, the grid structure is determined. According to data seen in Fig. 4.22, after 5.2 million elements, the results do not change significantly.

After grid structure is determined, governing equations should be included in the analysis. Governing equation is included in the analysis according to the physical problem. For instance, if there is a phase change in the problem, a proper differential equation set is included in the numerical analysis. Many CFD and multiphysics software have an extensive physic library. If the software does not have a valid equation set, some software allows adding equations manually. The initial and boundary

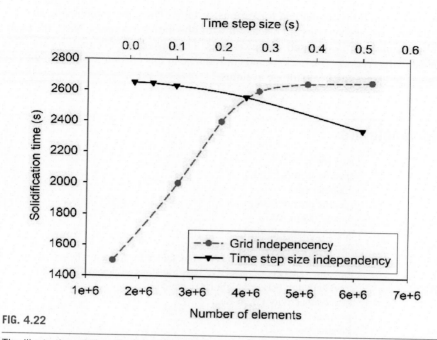

FIG. 4.22

The illustration of the effects of grid independency and time step size independency.

Adapted from Ref. [7].

conditions are applied to the numerical model. After this stage, the analysis is run. The analysis continues until the solution converges. The convergence criterion is either solution precision or the number of iterations. If the analysis is time dependent, the time step size independency is applied in this stage. The time step size is reduced until the result does not change significantly, as shown in Fig. 4.22. When the analysis is completed, the results are illustrated in the stage of postprocessing. It can be used for the postprocessing of either a separate software or a module integrated into the main software.

The most significant advantage of CFD analysis over experimental and analytical methods is to give rapid and precise results of the effect of many different designs and working conditions on the system. Therefore, they are commonly used for improving systems and processes. Here, the basics and procedure of the numerical analysis are introduced to the reader. Various illustrative CFD analysis examples from heat storage applications are introduced to the readers in the case studies.

4.9 Closing Remarks

In this chapter, first, the basics of the thermodynamic analysis based on the mass, energy, entropy, and exergy balances are introduced. Energy and exergy efficiencies and COP equations are defined for various systems. Then, thermodynamic analyses of the heat storage systems are discussed with illustrative examples. Next, the basics of heat transfer are expressed with examples. The basics of CFD analysis are introduced to the readers. The backgrounds introduced in this chapter are used in the case studies to assess the heat storage systems.

Nomenclature

A	area (m^2)
c	specific heat (kJ/kg K)
D	diameter (m)
e	specific energy (kJ/kg)
E	energy (kJ or kWh)
ex	specific exergy (kJ/kg)
Ex	exergy (kJ and kWh)
F	volume fraction of fluid or the view factor for radiation heat transfer
h	enthalpy (kJ/kg) or convective heat transfer coefficient (W/m^2 K)
H	total enthalpy (kJ) or height (m)
k	heat conductivity (W/m K)
K	permeability
KE	kinetic energy (kJ)
L	length (m)
m	mass (kg)
P	pressure (kPa)
PE	potential energy (kJ)
q	heat transfer per area (kJ/m^2)
Q	heat transfer (kJ or kWh)
R	thermal resistance (kW/m^2K)
s	entropy (kJ/kg)
S	total entropy (kJ)
t	time (s or h)
T	temperature (K or °C)
u	internal energy (kJ/kg)
U	total internal energy (kJ)
V	volume (m^3)
x, X	position (m)
$\dot{E}$	rate of energy (kW)
$\dot{Ex}$	rate of exergy (kW)
$\dot{m}$	mass flow rate (kg/s)
$\dot{q}$	heat transfer rate per area (kW/m^2)
$\dot{Q}$	rate of heat transfer (kW)
$\dot{S}$	rate of total entropy (kW)
$\dot{V}$	volumetric flow rate (m^3/s)
$\dot{W}$	rate of work (kW)

Greek Letters

α	absorptivity
ε	emissivity
η	energy efficiency
μ	dynamic viscosity (Ns/m^2)
ν	specific volume (m^3/kg) or kinematic viscosity (m^2/s)
ρ	density (kg/m^3) or reflectivity
σ	Stephan–Boltzmann constant
τ	transmissivity
ψ	exergy efficiency
φ	porosity

Subscripts

0	reference conditions
ave	average
b	boundary
boiler	boiler
bottom	bottom of the tank
bulk	bulk temperature of fluid
cap	capsule
ch	charging period
comp	compressor
con	condenser
dest	destruction
disch	discharging period
e	electricity or equivalent
env	environment
eva	evaporator
ex	exergy
f	fluid
final	final
flux	heat flux
gain	gain
gen	generation
HTF	heat transfer fluid
hws	hot water storage tank
ice	phase of ice
in, inlet	inlet
initial, init	initial
is	isentropic
loss	loss

m	fully mixed
mantle	mantle
net	net
out	outlet
ove	overall
p	constant pressure
pcm	phase change material
pump	pump
R	refrigeration
s	source or solid
sb	solid block
sen	sensible
sf	solid/fluid phase changing
sh	shaft
sm	storage medium
st	storing period
sur	surface
sys	system
t=0	initial conditions of any period
tank	storage tank
top	top of the tank
tur, turbine	turbine
valve	valve

References

[1] I. Dincer, Thermodynamics: A Smart Approach, first ed., John Wiley & Sons, Ltd., West Sussex, 2020.

[2] I. Dincer, M.A. Rosen, Thermal Energy Storage: Systems and Applications, second ed., 2010, https://doi.org/10.1002/9780470970751.

[3] D. Erdemir, Determination of effect of bottle arrangement in the sensible thermal energy storage system consisting of water-filled PET bottles on thermal performance, Hittite J. Sci. Eng. 6 (2019) 235–242, https://doi.org/10.17350/HJSE19030000153.

[4] J. Holman, Heat Transfer, McGraw-Hill, Newyork, 2014.

[5] E. Grimson, Correlation and utilization of new data on flow resistance and heat transfer for cross flow of gases over tube banks, Trans. ASME 59 (1937) 583–594.

[6] A. Bejan, Convection Heat Transfer, John Wiley & Sons, Inc., Hoboken, New Jersey, 2013.

[7] D. Erdemir, Numerical investigation of thermal performance of geometrically modified spherical ice capsules during the discharging period, Int. J. Energy Res. 43 (2019), https://doi.org/10.1002/er.4585.

Case Studies

5.1 Introduction

Heat storage systems have gained significant importance during the past two decades due to some key requirements, such as efficiency improvement, energy and cost savings, emissions reduction, and resilience increase. There has been increasing interest in various countries to promote the energy storage applications to offset the mismatch between demand and supply, especially for commercial and public buildings and numerous building complexes in various communities. In conjunction with this, several local governments have initiated rebate programs and incentives for increased utilization of such systems. These studies offer unique examples for people.

This chapter introduces numerous case studies of the building applications of heat storage systems in addition to the systems and illustrative examples presented in previous chapters. In the case studies, heat storage systems integrated into building energy systems are investigated though in many ways. Comprehensive modeling and analyses for the building applications of heat storage systems are discussed with the practical applications. The following methods are used in modeling and analysis of the heat storage systems used in the buildings: thermodynamic analysis, heat transfer analysis, computational fluid dynamics (CFD) modeling, and cost assessment.

Various heat storage systems driven or integrated with different energy sources are firstly evaluated through the fundamentals of thermodynamics based on energy and exergy approaches. In addition, heat transfer applications of heat storage systems are introduced. This is, furthermore, followed by the CFD studies of various heat storage systems.

5.2 Case Study 1: Sensible Heat Storage—Integrated Solar Heating System for a Fieldhouse

Solar heating implementation is considered one of the most common and mature renewable energy applications. The most significant problem with solar energy systems is not always available. Solar energy should be storage for night or winter uses. Heat storage systems are one of the most convenient to store solar energy, especially for heating purposes, as the type of the stored energy is the same with the type of energy demand. Here, in the present case study, a cost-effective heat storage system integrated into solar heating systems used in a fieldhouse is investigated. The system is introduced and evaluated experimentally based on the materials and data in Refs. [1–4].

5.2.1 System Description

The schematic diagram of the system is shown in Fig. 5.1. The system mainly consists of a solar air collector, fans, heat storage tank, and water-filled PET bottles. The view of the solar air collector is shown in Fig. 5.2a. The solar collector has been placed on the south side of the fieldhouse building. The total surface area of the solar collector is 320 m^2. The fieldhouse has 1650 m^2 of the total ground area and 13,200 m^3 of the closed volume. The peak heating load of the fieldhouse is 672 kW, respectively.

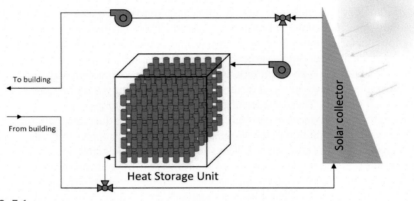

FIG. 5.1

The schematic view of the sensible heat storage—integrated solar heating system.

FIG. 5.2

(a) The view of the solar collector and fieldhouse, and (b) the view of the sensible heat storage tank consisting of PET bottles.

Adapted from Refs. [1,2].

To store solar energy for night use, the sensible heat storage unit that consists of 2560 water-filled PET bottles of 1.5-L volume is integrated into the system. The view of the heat storage unit is given in Fig. 5.2b. The dimensions of the storage unit are $4.5 \times 2 \times 2$ m. The diameter and height of the PET bottle are 50 and 300 mm, respectively. The total volume of the storage medium is 7680 m^3. The bottles are placed 16×40 in-line arrangement, and there are four rows. The heat storage tank is insulated with 60-mm rockwool.

The system primarily consists of three successive modes:

1) Normal heating: The air heated in the solar collectors is sent to the fieldhouse. It is also storing period.
2) Charging period: The air heated in the solar collectors is sent to the storage unit to charge the energy into the water.
3) Discharging period: The heating is performed with the stored energy by forcing the air through the PET bottles.

The modes and volumetric flowrates are controlled with PLC-based control system. The flowrate during the normal heating and charging periods is 22,500 m^3/h, and it is 10,000 m^3/h in the discharging period. The fieldhouse is open between 7:00 and 22:00. It is aimed to maintain the indoor temperature at 15°C. In this case study, the experimental observations and thermodynamic analysis are introduced. The heat transfer analysis of the heat storage unit is also discussed in the illustrative examples.

5.2.2 Thermodynamic Analysis

To assess the system, thermodynamic analysis based on energy and exergy calculation is applied to the system. The following assumptions are applied to the analysis to reduce the complexity of the analysis:

- The air is frictionless, and there are no pressure losses in the ducts and system elements.
- There is no heat loss in the ducts and system elements.
- Thermophysical features of the air are temperature dependent.
- The thermophysical properties of water remain unchanged with the temperature.

Although these assumptions tend to increase the energy and exergy efficiencies, but not their relative difference values. Thus, the results provide convenient information to evaluate the system under different operating conditions and design criteria. The schematic diagrams along with the respective notations of the operating periods are illustrated in Fig. 5.3.

The energy balance equation for the system is written as

$$(Energy\ input) - [(Energy\ recovered) + (Heat\ loss)] = Energy\ accumulation \quad (5.1)$$

Eq. (5.1) can be written as follows, by considering Fig. 5.3:

$$\left(\dot{E}_a - \dot{E}_b\right) - \left[\left(\dot{E}_d - \dot{E}_c\right) + \dot{Q}_{loss}\right] = \Delta E_{sys} \quad (5.2)$$

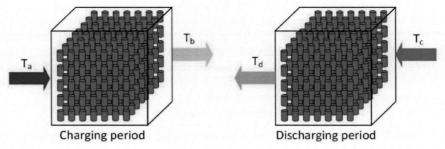

Charging period   Discharging period

FIG. 5.3

The schematic view of the operating principles of the system.

The first term on the left side of Eq. (5.2) is the charged energy. The first term in the bracket is the discharged energy or useful output. ΔE is the energy accumulation in the storage medium, and it is zero as the system works cyclic ($\Delta E = 0$).

The charged and discharged energies can be calculated, respectively, as

$$\dot{Q}_{ch} = \left(\dot{E}_a - \dot{E}_b \right) = \dot{m}_{ch} c_{p,ch} (T_a - T_b) \tag{5.3}$$

$$\dot{Q}_{disch} = \left(\dot{E}_d - \dot{E}_c \right) = \dot{m}_{disch} c_{p,disch} (T_d - T_c) \tag{5.4}$$

where $\dot{m}_{ch}$ and $\dot{m}_{disch}$ are the mass flow rates of the air in the charging and discharging period. $c_{p,ch}$ and $c_{p,disch}$ are the specific heats of the air at the bulk temperature of air during the charging and discharging period.

Heat loss from the storage tank to environment is calculated as

$$\dot{Q}_{loss} = \frac{\Delta T * A}{R_T} = \frac{\left(T_{storage} - T_{env} \right) * A}{R_T} \tag{5.5}$$

Here, $T_{storage}$ and T_{env} are the temperature of water stored and the environmental temperature, respectively. A is the total surface area of the heat storage unit.

The exergy balance equation for the system is written as

$$(Exergy\ input) - [(Exergy\ recovered) + (Heat\ loss)] - Exergy\ destruction$$
$$= Exergy\ accumulation \tag{5.6}$$

Eq. (5.6) can be written as the follows by considering the system notation:

$$\left(\dot{E}x_a - \dot{E}x_b \right) - \left[\left(\dot{E}x_d - \dot{E}x_c \right) + \dot{E}x_{\dot{Q}_{loss}} \right] - \dot{E}x_{dest} = \Delta Ex_{sys} \tag{5.7}$$

Here, $\left(\dot{E}x_a - \dot{E}x_b \right)$ is the charged exergy and $\left(\dot{E}x_d - \dot{E}x_c \right)$ is the discharged exergy. $\dot{E}x_{\dot{Q}_{loss}}$ denotes the exergy loss due to the heat loss from the storage tank to environment. $\dot{E}x_{dest}$ represents the exergy destruction in the system. Similar to energy accumulation in the system, the exergy accumulation in the system (ΔEx_{sys}) is zero as the system operates cyclic.

The charged and discharged exergies can be calculated, respectively, as

$$\dot{Ex}_{ch} = \dot{Ex}_a - \dot{Ex}_b = \dot{m}_{ch}c_{p,ch}\left[(T_a - T_b) - T_{env}\ln\left(\frac{T_a}{T_b}\right)\right] \tag{5.8}$$

$$\dot{Ex}_{disch} = \dot{Ex}_d - \dot{Ex}_c = \dot{m}_{disch}c_{p,disch}\left[(T_d - T_c) - T_{env}\ln\left(\frac{T_d}{T_c}\right)\right] \tag{5.9}$$

The exergy loss due to the heat loss is calculated as

$$\dot{Ex}_{\dot{Q}_{loss}} = \dot{Q}_{loss}\left[1 - \frac{T_{env}}{T_{storage}}\right] \tag{5.10}$$

The overall energy efficiency of the system is defined as

$$\eta = \frac{Total\ discharged\ energy}{Total\ charged\ energy} = \frac{\sum_{j=1}^{24}\left(\dot{E}_d - \dot{E}_c\right)_j}{\sum_{j=1}^{24}\left(\dot{E}_a - \dot{E}_b\right)_j} \tag{5.11}$$

The energy efficiencies in the charging and discharging periods can be defined, respectively, as follows:

$$\eta_{ch} = \frac{\Delta\dot{E}_{ch}}{\dot{E}_a - \dot{E}_b} = \frac{\left(\dot{E}_a - \dot{E}_b\right) - \dot{Q}_{loss}}{\dot{E}_a - \dot{E}_b} \tag{5.12}$$

$$\eta_{dis} = \frac{\dot{E}_d - \dot{E}_c}{\Delta\dot{E}_{dis}} = \frac{\left(\dot{E}_d - \dot{E}_c\right)}{\left(\dot{E}_d - \dot{E}_c\right) - \dot{Q}_{loss}} \tag{5.13}$$

The overall exergy efficiency of the system is defined as

$$\Psi = \frac{Exergy\ recovered}{Exergy\ input} = \frac{\sum_{j=1}^{24}\left(\dot{Ex}_d - \dot{Ex}_c\right)_j}{\sum_{j=1}^{24}\left(\dot{Ex}_a - \dot{Ex}_b\right)_j} \tag{5.14}$$

The exergy efficiencies in the charging and discharging periods can be defined, respectively, as follows:

$$\Psi_{ch} = \frac{\Delta\dot{Ex}_{ch}}{\dot{Ex}_a - \dot{Ex}_b} = \frac{\left(\dot{Ex}_a - \dot{Ex}_b\right) - \dot{Ex}_{\dot{Q}_{loss}} - \dot{Ex}_{dest}}{\dot{Ex}_a - \dot{Ex}_b} \tag{5.15}$$

$$\Psi_{dis} = \frac{\dot{Ex}_d - \dot{Ex}_c}{\Delta\dot{Ex}_{disch}} = \frac{\left(\dot{Ex}_d - \dot{Ex}_c\right)}{\left(\dot{Ex}_d - \dot{Ex}_c\right) - \dot{Ex}_{\dot{Q}_{loss}} - \dot{Ex}_{dest}} \tag{5.16}$$

5.2.3 Results and Discussion of Thermodynamic Analysis

The energy and exergy efficiencies are calculated with the data obtained during October and November. The monthly average of data for October and November is used in the calculations. Between 7:00 and 17:00, solar energy is used for the field-house heating and the energy charging. After 17:00, the fieldhouse is heated with the stored energy. Indoor, outdoor and storage tank temperatures, and solar radiation changes for October and November are shown in Fig. 5.4. The changes of T_a, T_b, T_c, and T_d with time are shown in Fig. 5.5.

In the light of the data presented in Figs. 5.4 and 5.5, the energy and exergy calculations are performed. Energy and exergy efficiencies and exergy destruction rates for October and November are given in Tables 5.1 and 5.2, respectively. As seen from energy efficiency values, the system seems to work as almost ideal as they change between 98% and 100%. On the other hand, the exergy efficiency values

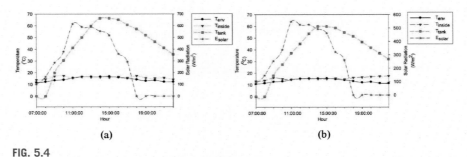

(a)   (b)

FIG. 5.4

Indoor, outdoor, and storage tank temperatures and solar radiation changes for (a) October and (b) November.

Data from Ref. [2].

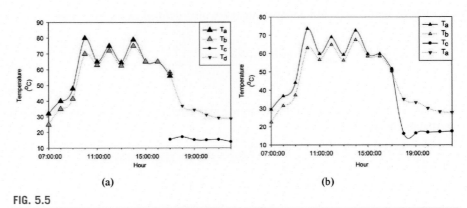

(a)   (b)

FIG. 5.5

T_a, T_b, T_c, and T_d values for (a) October and (b) November.

Data from Ref. [2].

Table 5.1 Energy and Exergy Efficiencies and Exergy Destruction Rates for October.

Hour	η	Ψ	$\dot{Ex}_{dest}$ (kW)
7:00—8:00	0.99	0.52	30,525
8:00—9:00	0.99	0.75	24,008
9:00—10:00	0.99	0.74	31,588
10:00—11:00	0.99	0.80	51,458
11:00—12:00	0.98	0.94	10,270
12:00—13:00	0.98	0.93	14,637
13:00—14:00	0.98	0.94	9546
14:00—15:00	0.98	0.91	19,575
15:00—16:00	0.98	0.92	14,432
16:00—17:00	0.98	0.95	9702
17:00—18:00	0.98	0.94	8708
18:00—19:00	0.99	0.03	29,851
19:00—20:00	0.99	0.07	28,804
20:00—21:00	0.99	0.09	21,105
21:00—22:00	0.99	0.16	19,326
22:00—23:00	0.99	0.18	18,208

Table 5.2 Energy and Exergy Efficiencies and Exergy Destruction Rates for November.

Hour	η	Ψ	$\dot{Ex}_{dest}$ (kW)
7:00—8:00	1.00	0.48	30,083
8:00—9:00	1.00	0.71	25,481
9:00—10:00	1.00	0.70	33,231
10:00—11:00	1.00	0.77	55,436
11:00—12:00	0.99	0.90	15,891
12:00—13:00	0.99	0.89	20,896
13:00—14:00	0.99	0.90	15,015
14:00—15:00	0.99	0.88	25,832
15:00—16:00	0.97	0.96	6218
16:00—17:00	0.97	0.96	6321
17:00—18:00	0.98	0.96	5405
18:00—19:00	1.00	0.03	29,019
19:00—20:00	1.00	0.07	27,732
20:00—21:00	1.00	0.10	19,995
21:00—22:00	1.00	0.17	17,988
22:00—23:00	1.00	0.20	16,843

show how the system works far away from the ideal since they change between 3% and 96%. The exergy efficiency values are lower than energy efficiency values, as exergy calculations are taken into consideration the irreversibilities inside the system boundary, in addition to heat loss crossing the system boundary. While the average values of energy and exergy efficiencies for the charging period are 99% and 80% respectively, they are 94% and 3% for the discharging period. The main reason for the difference between the values in charging and discharging periods is the difference in the temperature between environmental and working fluid temperatures.

The energy efficiency in the sensible heat storage—integrated solar heating system is the ratio of useful heat output to the charged heat. On the other hand, the exergy efficiency includes increasing thermodynamic unavailability, which is defined as exergy destruction. Also, the exergy efficiency provides a measure of how the entire system and process approach ideality, while energy efficiencies do not. The overall energy and exergy efficiency values are calculated as 79.85% and 51.89% for October and 69.95% and 46.26% for November. The variation between the months studied occurs due to the environmental temperature. Here, it should be noted that exergy efficiency has the capability to define the potential of the system according to reference conditions, too. When the temperature differences between the working temperature of the system and the environmental temperature increase, the exergy efficiency tends to decrease. Therefore, the exergy efficiency values of October are higher than November. This situation is seen in the exergy destruction values given in Tables 5.1 and 5.2.

5.2.4 Heat Transfer Analysis for the Heat Storage Unit Consisting of PET Bottles

To analyze the system performance and determine the optimum design, heat transfer analysis should be performed to the heat storage unit. The heat storage unit consists of water-filled PET bottles. PET bottles act as the capsules, and the air flows over them. Therefore, the arrangement of the bottles and the volume of the bottles have a great impact on the performance of the system. It can be modeled by using the background of heat transfer analysis of the external flow over the tube bank. The modeling of the heat storage unit consists of water-filled PET bottles as a tube bank is studied in Ref. [1].

The selected control volume has a 10×10 arrangement as illustrated in Fig. 5.6. As the bottles are placed layer by layer on top of each other, per bottle line can be assumed as a tube. So, the entire storage pack can be assumed as the tube bank. A comprehensive illustrative example has been discussed in Example 7 in Chapter 4. The analysis made here is performed based on the equations in Example 4.7. Here, the results are discussed for different bottles and arrangements. Fig. 5.7 shows the view of the PET bottles studied, which are 0.5, 1.5, and 5 L. When the heat transfer analysis is considered for the heat transfer, only the bottle diameter has an impact on the analysis. The height of the bottle changes the heat transfer surface area. S_N is the

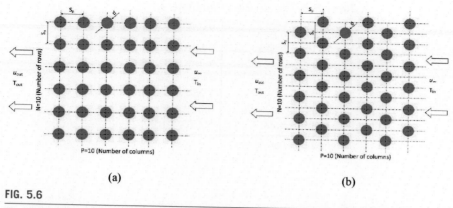

FIG. 5.6

The PET bottle arrangement in the storage tank (a) in-line mode and (b) staggered mode.

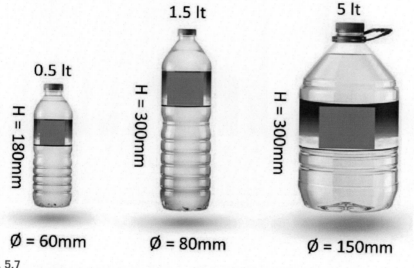

FIG. 5.7

The view of the PET bottles studied.

Adapted from Ref. [1].

distance between the bottles at the plane normal to the airflow, and S_P is the distance between the bottles at the plane parallel to the airflow. Those two parameters have a significant impact on the performance of the system. In the analysis, to compare the results properly, the total volume of stored water is taken as constant, 1500 L. The number of layers is changed to set the total water volume of 1500 L.

The studied parameters in the work are given in Table 5.3. Base on the parameters given in Table 5.3, first, the amount of the stored energy is calculated. The

Table 5.3 The Studied Parameters.

Parameter	Value
Bottle volume, liter	0.5, 1.5, 5
Arrangements	In-line, staggered
S_N	1.25d, 1.5d, 2d
S_P	1.25d, 1.5d, 2d
Inlet velocity, u_∞, m/s	0.5, 1, 1.5, ..., 9, 9.5, 10
Inlet temperature, T_{in},°C	50, 70, 80, 90

Modified from Ref. [1].

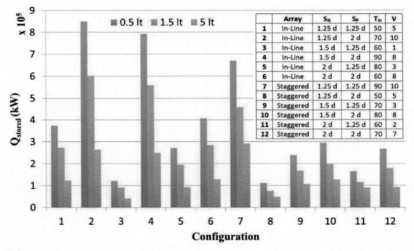

FIG. 5.8

The amount of the stored heat for the selected configurations.

Modified from Ref. [1].

amount of the stored energy is shown in Fig. 5.8 for the selected configurations. The amount of the stored energy decreases with the increasing bottle volume. The diameter of the bottle changes with the bottle volume. The changes in the diameter affect the heat transfer surface area and the heat transfer coefficient between the bottle and airflow. A higher diameter provides a higher heat transfer coefficient, as the maximum velocity and Reynolds number increase with the increasing diameter. Then, a higher Reynolds number causes a higher Nusselt number and heat transfer coefficient. On the other hand, the total heat transfer surface area decreases with the increasing bottle volume. That is why the stored energy decreases with the increasing bottle volume.

The effects of the design parameter, which are S_N and S_P on the amount of the stored heat, are demonstrated in Fig. 5.9. In the in-line arrangement, the stored energy amount decreases with the increasing S_N, as the turbulent effect decreases with the increasing S_N. Otherwise, there is no significant effect of S_P on the thermal performance of the stored heat. In the staggered arrangement, in contrast with the in-line arrangement, the amount of stored heat increases with increasing S_N. Consequently, to maintain a higher stored heat in a staggered arrangement, S_N should be higher, and S_P should be lower.

5.2.5 Closure

The following key findings are summarized out of this case study:

- The results indicate that water-filled PET bottles are an effective way to storage medium because PET bottles act as heat storage capsules that work as heat exchanger and storage material container in the solar energy heating system.
- The overall energy and exergy efficiencies have been calculated as 79.85% and 51.89% for October. These values are 69.95% and 46.26% for November, respectively. The exergy efficiency is lower during the discharging, because of higher temperature difference. Lower exergy efficiency values can be enhanced by modifying operating conditions. The charging and discharging performance of the system can also be improved with PET bottle arrangements and sizes.
- There is no significant difference between in-line and staggered arrangements in terms of the stored energy amount in the higher Reynolds number.
- While lower S_N and S_P should be preferred in the in-line arrangement, higher S_N and lower S_P should be used in the staggered arrangement.

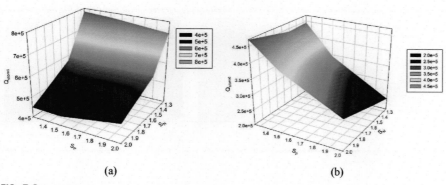

(a)   (b)

FIG. 5.9

The effects of S_N and S_P on the amount of stored heat for (a) in-line and (b) staggered arrangements for the case of $u_\infty = 5$ m/s, $T_{in} = 90°C$ and 0.5-L PET bottle.

Modified from Ref. [1].

5.3 Case Study 2: Heat Storage in a Vertical Hot Water Tank and Thermal Stratification

Hot water tanks are one of the most common heat storage applications in the buildings, as they are used for meeting the hot water, which is one of the most used demand items in daily life. They are an essential part of the solar domestic hot water systems. Thermal stratification is the most significant parameter to measure the performance of heat storage in the hot water tanks. Therefore, to enhance the degree of thermal stratification, many studies investigating the effect of the tank design and operating conditions and various devices have been introduced in the literature and the sector. In this case study, the effect of the thermal stratification on the energy and exergy efficiencies is discussed with the experimental observation in Refs. [5,6].

5.3.1 System Description

In the case study, a vertical hot water tank is studied experimentally, shown in Fig. 5.10. A mantled hot water tank, which is commonly used in solar domestic hot water system, is studied. The tank is 200 L of the effective storage volume. The diameter and height of the tank are 51 and 77.5 cm, respectively. Mantle gap is 15 mm. The tank consists of two sections, which are cold and hot sections. In the cold-water section, the main line water is kept temporarily. The cold and hot water sections are connected each other with a pipe, as named cold water pipe in Fig. 5.10. The cold water as much as the volume of hot water consumed enters the hot water zone from the cold-water zone. The cold-water zone fills from the mains. The hot water is discharged from the top of the hot water section with a pipe, named the hot water pipe in Fig. 5.10. The temperature measurement is gathered at the center of the tank in every 100 mm from the tank bottom. To enhance the

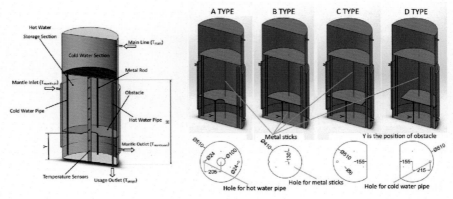

FIG. 5.10

The schematic view of the tank and the obstacles placed inside the tank.

Adapted from Refs. [5,6].

thermal stratification, four different obstacles are placed inside the tank in different positions from the tank bottom. The obstacles keep the cold water coming from the cold-water section under of it. Thus, the amount of the hot water increases. The details of the obstacles are shown in Fig. 5.10. The effect of the obstacle on the degree of the thermal stratification in the tank is determined.

To evaluate the thermal stratification in a hot water tank, first, the temperature distribution on the tank is examined. It is expected to be minimum the cold-water volume in the bottom of the tank, while the hot water volume is desired to the maximum. However, the temperature distribution does not provide a clear comparison for different cases. Therefore, the energy and exergy contents of the tank should be determined considering the thermal stratification of the tank. The background related to the energy and exergy analyses of the thermally stratified heat storage tank is introduced in Section 4.6. Here, the energy and exergy efficiency values are calculated by using equations in Section 4.6.

5.3.2 Results and Discussion

To apply the thermodynamic analysis for hot water, it is required to know the temperature distribution inside the tank. The temperature distribution can be determined experimentally, numerically, or theoretically. When the temperature distribution is known, energy and exergy calculations are performed by using the equations in Section 4.6. In this case study, four obstacles are placed inside the tank to improve the temperature distribution in different positions from the tank bottom. The variation of the temperature gradient with the time is shown in Fig. 5.11. The temperatures are gathered every 10 cm from the tank bottom. It is clear from Fig. 5.11 that the thermal stratification starts to develop once energy is charged. After an hour, there are steady thermal stratification conditions inside the tank. The water flow rates in the mantle (primary line) and mains (secondary line) are 4.5 L/min. The main objective of placing obstacles inside the tank is to force to keep cold water coming from the mains under the obstacle. Thus, the cold water volume is minimized, and the volume of hot water increases significantly. Fig. 5.11 also shows how the obstacles work to improve the heat storage performance of the tank. It is clear that the obstacles achieve to limit the cold water to reach the upper part of the tank.

To observe how the obstacles affect temperature distribution according to the ordinary tank that is without obstacles, the temperature distributions for the case with obstacle should be compared with the temperature distribution of the ordinary tank. Fig. 5.12 illustrates the comparison of the temperature distributions for the case without obstacle. It is clear from Fig. 5.12 that the obstacles improve the thermal stratification. The optimum obstacle position is observed in between 200 and 300 mm, according to the temperature distribution. The studied obstacle geometries are considered; obstacle A provides the best heat storage performance.

The energy and exergy contents of the tank at the end of 120 min test period are calculated by using the temperature distribution shown in Fig. 5.12. The continuous-linear temperature distribution model is applied in the thermodynamic analysis as

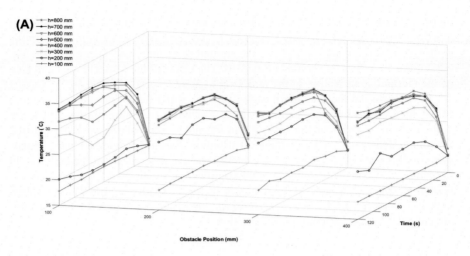

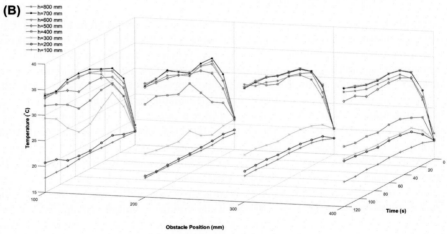

FIG. 5.11

The temperature distribution change with the time in the tank for (a) obstacle A, (b) obstacle B, (c) obstacle C, and (d) obstacle D.

Adapted from Ref. [5].

the temperature measurements are taken from the different heights inside the tank. In this model, it is assumed that the temperature changes linearly between two temperature measurement points (for detail information, please see Section 4.6). Energy and exergy contents of the ordinary tank are calculated to be 2167 and 554 kJ, respectively. The variation of the energy and exergy contents with the obstacle position is shown in Fig. 5.13. The change in energy content is lower than the exergy content. Therefore, exergy content provides more comparable information about the

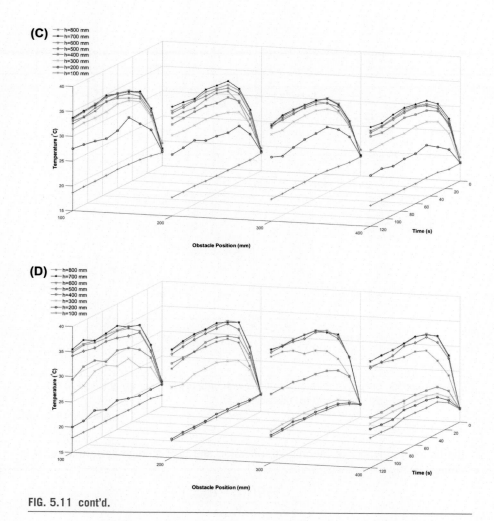

FIG. 5.11 cont'd.

effect of the obstacles and their positions. It is clear from Fig. 5.13 that obstacle A provides the highest increase in the energy and exergy content. According to energy and exergy contents, the obstacle should be positioned between 200 and 300 mm from the tank bottom.

In addition to the energy and exergy calculations, some dimensionless numbers are used for determining the degree of the thermal stratification (see Fig. 2.9). Richardson number is one of the most used dimensionless numbers to evaluate the degree of the thermal stratification. Richardson number defines the flow and stratification in the thermally stratified heat storage units. It governs the ratio of buoyancy forces to mixing forces. The higher Richardson number values mean

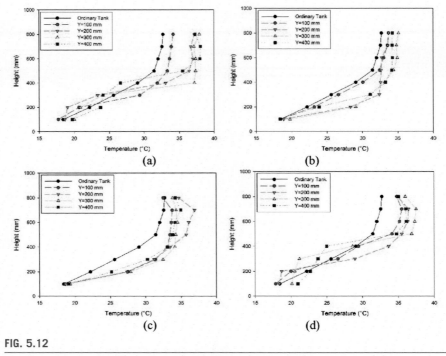

FIG. 5.12

The comparison of temperature distributions with ordinary tank for (a) obstacle A, (b) obstacle B, (c) obstacle C, and (d) obstacle D.

Adapted from Ref. [5].

the better the degree of thermal stratification and heat storage performance. It is defined as

$$Ri = \frac{g\beta H\left(T_{top} - T_{bottom}\right)}{V_{main}^2} \tag{5.17}$$

The variation of the Richardson number with the obstacle position is shown in Fig. 5.14. The characteristic of the Richardson number is different from the energy and exergy contents as it considers as only the top and bottom temperatures of the tank. The optimum obstacle position is different for each position. When obstacle A is considered, the optimum position of the obstacle is 300 mm. When the temperature distributions, energy and exergy contents, and Richardson number are considered simultaneously, it should be emphasized that the performance of the thermally stratified heat storage tanks should be evaluated in many ways. When all parameters are considered, the best heat storage is maintained by obstacle A in between 200 and 300 mm.

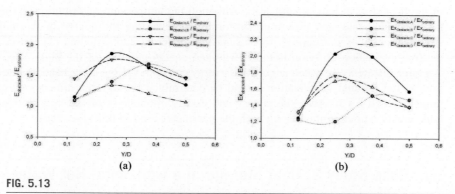

FIG. 5.13

The comparison of temperature distributions with ordinary tank for (a) obstacle A, (b) obstacle B, (c) obstacle C, and (d) obstacle D.

Adapted from Ref. [5].

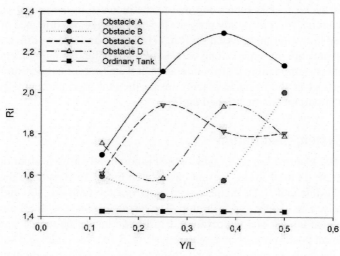

FIG. 5.14

The variation of Richardson number with the obstacle position.

5.3.3 Closure

Heat storage tanks are one of the most common and mature heat storage techniques, as they meet one of the most used demand items, hot water. They are also one of the most known energy storage methods of renewables, as they are used in the solar domestic hot water storage systems. Thermal stratification is the major parameter on the performance of the hot water tanks. It affects the energy and exergy content of the heat storage tank significantly. Although the temperature distribution in the tank provides easy and meaningful information regarding the degree of thermal stratification, it is not enough to evaluate the different cases studied. Therefore, the first evaluation criterion is the energy and exergy contents of the tank. There is a huge difference between energy and exergy contents. On the other hand, the degree of the thermal stratification should be determined through the dimensionless numbers used for it. Also, the energy and exergy efficiencies of the heat storage tank should be determined along with the parameter used in this case study.

5.4 Case Study 3: Heat Storage in a Horizontal Hot Water Tank

Although there are many classification methods for hot water tanks, the most common one is related to their position, which is vertical and horizontal. The thermal performance and thermal stratification of the hot water tank are generally better than the horizontal one as the vertical distance is higher. However, the use of the horizontal hot water tank is increasing due to the increase in the use of vacuum tube solar collectors. This case study presents the heat storage performance of a horizontal mantled hot water tank, based on Ref. [7]. Energy and exergy efficiencies of the horizontal tank are presented along with additional evaluation parameters, which are dimensionless temperature distribution and hot water output temperature. Additionally, the effect of the obstacle placing inside the tank is experimentally discussed through the parameters mentioned earlier.

5.4.1 System Description

A horizontal mantled hot water tank, which is commonly used in solar domestic hot water system, is used in the present case study. Fig. 5.15 illustrates the schematic view of the test tank. The diameter and length of the tank are 400 and 1000 mm, respectively. The mantle gap is 20 mm. The test is insulated with 80-mm rockwool to reduce heat losses. The temperature distribution inside the tank is gathered from nine different positions as shown with red in Fig. 5.15. Also, inlet and outlet temperatures of the tank are measured. Two obstacles are placed perpendicular to the flow direction to enhance the thermal performance of the tank. The effect of the positions of the obstacles (a and b) on the heat storage performance is researched. In the case study, mantle and main flow rates are 4 and 4.5 L/min, respectively. Mantle inlet temperature (T_1) is 60°C, and the mains inlet temperature (T_2) is 17°C. The reference temperature is 20°C, respectively.

5.4.2 Thermodynamic Analysis of Horizontal Mantled Hot Water Tank

Thermodynamic balance equations, which are mass, energy, entropy, and exergy balances, for the mantle hot water tank can be written as

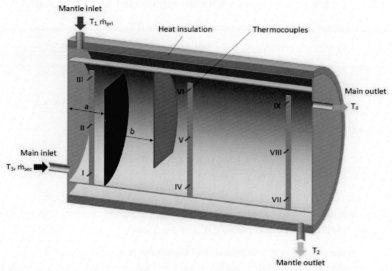

FIG. 5.15

The schematic view of the horizontal mantled hot water tank and the obstacles.

Modified from Ref. [7].

$$\dot{m}_1 = \dot{m}_2 = \dot{m}_{pri} \quad and \quad \dot{m}_3 = \dot{m}_4 = \dot{m}_{sec} \qquad (5.18)$$

$$\left[\dot{m}_1 h_1 + \dot{m}_3 h_3\right] - \left[\dot{m}_2 h_2 + \dot{m}_4 h_4\right] - \dot{Q}_{loss} = \Delta E_{sys} \qquad (5.19)$$

$$\left[\dot{m}_1 s_1 + \dot{m}_3 s_3\right] - \left[\dot{m}_2 s_2 + \dot{m}_4 s_4\right] + \dot{S}_{\dot{Q}_{loss}} = \Delta S_{sys} \qquad (5.20)$$

$$\left[\dot{m}_1 ex_1 + \dot{m}_3 ex_3\right] - \left[\dot{m}_2 ex_2 + \dot{m}_4 ex_4\right] - \dot{E}x_{\dot{Q}_{loss}} = \Delta Ex_{sys} \qquad (5.21)$$

General energy balance equation for the mantled hot water tank can be written as

$$Heat\ input - Heat\ recorvered - Heat\ loss = Heat\ accumulation \qquad (5.22)$$

Eq. (5.22) can be written as follows by considering Fig. 5.15:

$$\left[\dot{m}_{pri}(h_1 - h_2)\right] - \left[\dot{m}_{sec}(h_3 - h_4)\right] - \dot{Q}_{loss} = \Delta E_{sys} \qquad (5.23)$$

The energy efficiency can be calculated as

$$\eta = \frac{Heat\ recorvered}{Heat\ input} = \frac{\sum_{t=0}^{n}\left[\dot{m}_{sec}(h_3 - h_4)t\right]}{\sum_{t=0}^{n}\left[\dot{m}_{pri}(h_1 - h_2)t\right]} \qquad (5.24)$$

General exergy balance equation for the mantled hot water tank can be written as

Exergy input − Exergy recorvered − Exergy loss − Exergy destruction

$$= Exergy\ accumulation \qquad (5.25)$$

Eq. (5.25) can be written as follows by considering Fig. 5.15:

$$\left[\dot{m}_{pri}(ex_1 - ex_2) \right] - \left[\dot{m}_{sec}(ex_3 - ex_4) \right] - \dot{Ex}_{\dot{Q}_{loss}} = \Delta Ex_{sys} \qquad (5.26)$$

The energy efficiency can be calculated as

$$\psi = \frac{Exergy\ recorvered}{Exergy\ input} = \frac{\sum_{t=0}^{n} \left[\dot{m}_{sec}(ex_3 - ex_4)t \right]}{\sum_{t=0}^{n} \left[\dot{m}_{pri}(ex_1 - ex_2)t \right]} \qquad (5.27)$$

In the thermodynamic analysis, the following assumptions are applied to the hot water tank to reduce the complexity of the analysis.

- The working fluid is frictionless, and pumping power is zero.
- Thermophysical properties of the working fluid are temperature dependent.
- Heat loss from the tank is neglected.

5.4.3 Results and Discussion

Although there are many parameters to evaluate the heat storage performance of the vertical hot water tank, such as Richardson number, Stratification number, MIX number, and so on, there is no evaluation parameter for horizontal hot water tanks in the open literature. Therefore, the most common methods are used for evaluating the heat storage performance of the horizontal hot water tanks, which are energy and exergy efficiencies and temperature distribution of the tank.

The temperature distribution of the hot water tank is a fast and easy criterion to evaluate the heat storage performance. To normalize the temperature values in the tank, dimensionless temperature is also used. The dimensionless temperature can be calculated as

$$T^* = \frac{T_i - T_{min}}{T_{max} - T_{min}} \qquad (5.28)$$

where T_i is any temperature value inside tank. Here, it is the measured temperature value shown in Roman numerals in Fig. 5.15. For a hot water tank, T_{min} is the mains inlet temperature (T_3), and T_{max} is the primary-line inlet temperature (T_1). The dimensionless temperature distribution normalizes the temperature values in a range between 1 and 0. Thus, it provides more meaningful information to compare the different cases studied.

Fig. 5.16 shows the dimensionless temperature distribution inside the tank. When the ordinary tank without obstacle is considered, almost half of the tank has the cold water coming from the mains line and the hot water has stayed at top half of the tank. The reason occurring this is that the cold water coming from the mains easily mixes water and disturbs the thermal stratification in the tank. Therefore, both the water

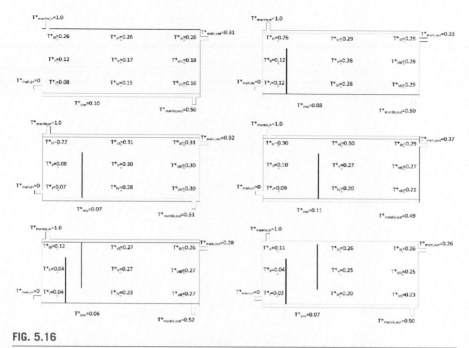

FIG. 5.16

The dimensionless temperature distribution for the different cases studied.

Adapted from Ref. [7].

temperature level and hot water level are lower than the other cases that are placed obstacles. It is clear from Fig. 5.16 that placing obstacle(s) vertical to flow direction prevents the blending of cold water with the stored hot water inside the tank easily because obstacle(s) act(s) as a flow and thermal barrier. Thus, the obstacle(s) provide(s) a higher volume water storage at a higher temperature. When a single obstacle is placed inside the tank, the cold water coming from the mains hits the obstacle and then goes upward. The cold water does not proceed in the axial direction. There is a hot water zone after the obstacle at a higher temperature than the ordinary tank. The obstacle should be placed closer to the mains inlet. When two obstacles are placed inside the tank, as indicated as A and B in Fig. 5.15, the second obstacle (obstacle B) does not change the temperature distribution and hot water volume significantly for the case of the first obstacle placed closer to the mains inlet. When the temperature distribution is taken into consideration, obstacle A is enough to provide the higher temperature and volume of stored water, when it is placed closer to the mains inlet.

As mentioned in previous case studies, although the temperature distribution provides significant information regarding the heat storage performance of the tank, it is not enough to evaluate it alone. Therefore, other performance criteria should be included in the evaluation procedure. The temperatures of primary and secondary outlets are one of those, as they directly affect the thermodynamic efficiencies. Of course, energy and exergy efficiencies are also essential evaluation criteria. Primary outlet temperature (mantle outlet) is important for the overall system and collector efficiencies. It

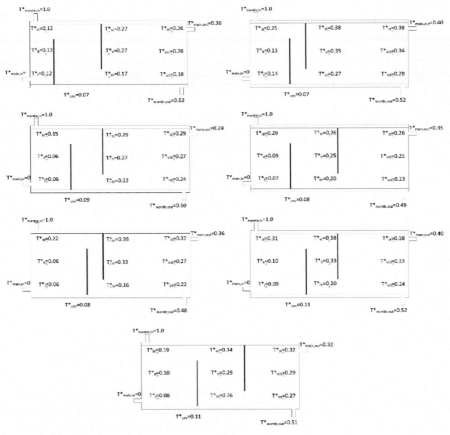

FIG. 5.16 cont'd.

should be minimized to achieve higher efficiencies. Also, lower mantle outlet temperature provides higher flow rates for the system working with natural circulation. Secondary outlet temperature (mains outlet or usage outlet) is the parameter that defines the user's satisfaction, as water is used at the secondary outlet conditions. It should be maximized in terms of both temperature and volume. The primary and secondary outlet temperatures and the energy and exergy efficiencies are listed in Table 5.4 for the studied cases. As seen from Table 5.4, while the obstacle(s) increase(s) the temperature of the usage water, the temperature of mantle outlet is reduced. Although there is no clear correlation between the number and position of obstacle and outlet temperatures, the best case is almost same with the cases having the better temperature distribution. The best case is seen when obstacle A is placed closer to mains inlet.

When the energy and exergy efficiencies are considered, they tend to increase with the obstacle placement. While the energy efficiencies change between 95.1% and 99.3%, the exergy efficiencies change between 8% and 19%. It is clear from those results that although energy efficiencies indicate the hot water tank works

Table 5.4 Mantle Outlet and Mains Outlet Temperature, and Energy and Exergy Efficiencies.

Case	$T_{mantle,out}$ (°C)	$T_{main,out}$ (°C)	η (%)	ψ (%)
Ordinary tank	40.4	30.1	95.11	8.01
$a = 150$ mm	38.9	33.5	97.22	19.23
$a = 250$ mm	39.9	30.7	99.06	17.36
$a = 350$ mm	38.7	33.4	97.84	15.46
$a = 150$ mm $b = 100$ mm	39.0	33.7	96.72	19.63
$a = 150$ mm $b = 200$ mm	38.6	30.4	95.68	17.21
$a = 150$ mm $b = 300$ mm	39.7	32.5	99.13	16.41
$a = 250$ mm $b = 100$ mm	39.8	32.4	98.08	17.42
$a = 250$ mm $b = 200$ mm	38.6	29.6	96.58	16.05
$a = 250$ mm $b = 300$ mm	38.9	32.0	97.01	13.42
$a = 350$ mm $b = 100$ mm	38.9	33.1	97.40	16.37
$a = 350$ mm $b = 200$ mm	39.2	33.7	98.93	14.27
$a = 350$ mm $b = 300$ mm	39.1	33.7	99.30	19.21

Modified from Ref. [7].

almost ideal, exergy efficiency values point to opposite. Energy and exergy efficiencies in the ordinary tank are nearly 95.1% and 8%, respectively. The highest exergy efficiency is seen in the case which obstacle A is placed closer to the mains inlet. The best energy and exergy efficiencies are not seen in the same cases, as the environmental temperatures measured during the experiments are different. When the temperature distribution, outlet temperatures, and thermodynamic efficiencies are assessed together, to enhance the heat storage performance of the horizontal hot water tank, obstacle A should be placed 200 mm from the mains inlet.

5.4.4 Closure

The following key results are observed from this case study:

- The horizontal hot water tanks are required to enhance to enhance their performance as the mains inlet easily mixes the stored water. Almost half of the bottom of the tank is under effect of the cold water coming from the mains.
- Obstacle placed vertical to flow direction is a cheap, easy-to-apply, and effective method to improve the performance of the horizontal hot water tanks.
- Obstacle is the flow and thermal barrier inside the tank and keeps the cold water behind it. Thus, the temperature level and hot water volume inside the tank increase significantly.
- Placing dual obstacles inside the tank does not change the heat storage performance. For the best performance, obstacle A should be placed at 200 mm from the mains inlet.

5.5 Case Study 4: Hybrid (Sensible and Latent) Heat Storage in Hot Water Storage Tank

In certain heat storage systems, sensible and latent heat storage mediums are used simultaneously (see Section 2.7.1) to increase the capacity of the heat storage. Two or more storage mediums are used in the heat storage unit. One of the most common applications of hybrid heat storage is to place phase change materials (PCMs) inside the hot water tank. In the present case study, two applications of hybrid heat storage in the hot water tank are introduced to the readers. The use of the PCMs inside the horizontal and vertical hot water tanks is studied with experimental observations.

5.5.1 System Description

In this case study, vertical and horizontal mantled hot water tanks are used as the test tanks. Experimental observations are obtained from Refs. [8–10]. The schematic views of the vertical and horizontal water tanks are shown in Fig. 5.17. While the vertical tank is 450 L, the horizontal tank is 200 L. Paraffin is used as the PCM and placed inside the tank with the cylindrical capsules. The volume of the capsules placed inside the vertical tank is 5 L and that placed inside the horizontal tank is 1 L. In order to obtain the temperature distribution inside the tank, thermocouples are placed inside the tanks, as shown in Fig. 5.17. Inlet and outlet temperatures are also measured. The number of the capsules is increased to determine the effect of the amount of paraffin on the heat storage performance.

To increase the capacity of the heat storage and hot water volume to be obtained, one of the most used methods is to place PCMs inside the tank. Paraffin is one of the most applicable PCMs in the hot water tank due to the following significant features:

- Adjustable melting point
- Higher latent heat
- Low cost

PCMs are placed inside the tank with capsules, tubes, or a single container. The method of packaging, the geometry of packaging, and their position inside the tank significantly affect the heat storage performance of the tank.

The main issue for placing PCMs inside the tank is to replace the hot water volume with the capsule volume. Namely, ready-to-use hot water volume decreases with the increasing capsule volume. Therefore, it should be optimized according to the tank design and operating conditions. The change in hot water volume with the capsule volume is shown in Fig. 5.18 for the vertical hot water tank. It is clear from Fig. 5.18 that the effective hot water volume decreases with the increasing the number of the paraffin capsules, although the capacity of the heat storage increases. This trend is similar for every hot water tanks.

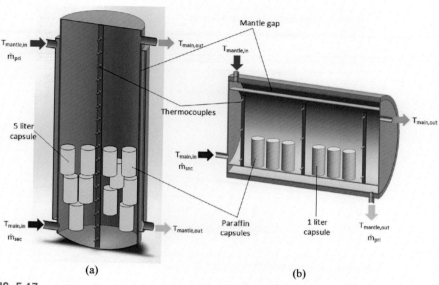

$T_{mantle,in}$
$\dot{m}_{pri}$
$T_{main,out}$
Mantle gap
$T_{mantle,in}$

Thermocouples

5 liter
capsule

$T_{main,in}$
$\dot{m}_{sec}$

$T_{main,out}$

$T_{main,in}$
$\dot{m}_{sec}$
$T_{mantle,out}$
Paraffin
capsules
1 liter
capsule
$T_{mantle,out}$
$\dot{m}_{pri}$

(a)   (b)

FIG. 5.17

The schematic view of the mantled hot water tanks: (a) vertical and (b) horizontal.

5.5.2 Results and Discussion

This case study aims to determine the effect of paraffin placed inside both the vertical and horizontal hot water tanks, which are commonly used in the hot water systems in the buildings. The most important criteria in a hot water tank for users are the hot water volume to be obtained and its temperature level. To increase the heat storage capacity and hot water output, it is a common method to place encapsulated paraffin inside the tank. As mentioned previously, the main issue for placing PCMs inside the tank is to replace the hot water volume with the capsule volume. Namely, ready-to-use hot water volume decreases with the increasing capsule volume. Therefore, it should be optimized according to the tank design and operating conditions.

For the vertical mantled hot water tank, Fig. 5.19 shows the total volume of hot water output for different flow rates for the case of $T_{mantle,in} = 80°C$. To determine the optimum paraffin amount, 4, 8, 12, and 16 capsules are placed in the tank, respectively. Although the hot water output volume increases with 4, 8, and 12 paraffin capsules, it reduced dramatically with 16 capsules. This situation occurs in all cases studied. To determine precisely the number of paraffin capsules, the experiments for 13 and 14 capsules have been conducted, respectively. While the hot water output increases with 13 capsules, it reduces with 14 capsules. Consequently, the optimum number of capsules is observed to be 13 for the 450 L tank, which has 5 L paraffin capsule that has 2.5 kg paraffin.

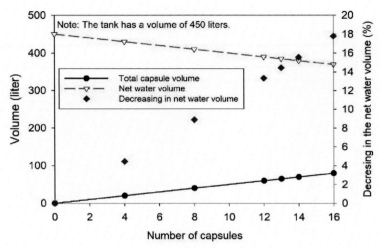

FIG. 5.18

The volume changes in the tank with the number of capsules for the vertical mantled hot water tank.

Data from Ref. [10].

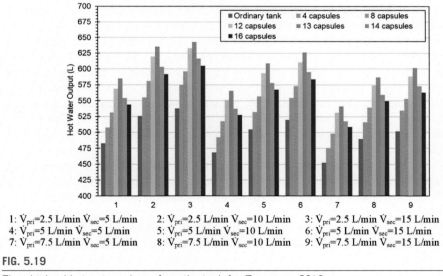

1: $\dot{V}_{pri}$=2.5 L/min $\dot{V}_{sec}$=5 L/min   2: $\dot{V}_{pri}$=2.5 L/min $\dot{V}_{sec}$=10 L/min   3: $\dot{V}_{pri}$=2.5 L/min $\dot{V}_{sec}$=15 L/min
4: $\dot{V}_{pri}$=5 L/min $\dot{V}_{sec}$=5 L/min   5: $\dot{V}_{pri}$=5 L/min $\dot{V}_{sec}$=10 L/min   6: $\dot{V}_{pri}$=5 L/min $\dot{V}_{sec}$=15 L/min
7: $\dot{V}_{pri}$=7.5 L/min $\dot{V}_{sec}$=5 L/min   8: $\dot{V}_{pri}$=7.5 L/min $\dot{V}_{sec}$=10 L/min   9: $\dot{V}_{pri}$=7.5 L/min $\dot{V}_{sec}$=15 L/min

FIG. 5.19

The obtained hot water volume from the tank for $T_{mantle,in} = 80°C$.

Data from Ref. [10].

The comparison of the energy and exergy efficiencies is shown in Table 5.5 for the ordinary tank, and the tank has 13 paraffin capsules. During the experiments, heat input is constant for all cases. Only, heat output (hot water output) changes due to the amount of paraffin. With increasing hot water amount, the energy and exergy efficiencies have increased as the useful output increases. Therefore, the highest efficiencies are seen in the cases in which 13 capsules are placed.

For the horizontal hot water tank, the hot water output variation with the number of capsules is shown in Fig. 5.20. The volume of hot water obtained from the tank increases with the increasing paraffin amount inside the tank up to four capsules. After the cases with five and six capsules, the hot water output volume decreases dramatically. When four capsules, containing 5 kg paraffin, are placed inside the tank, the hot water output has increased by 40%. When the vertical and horizontal hot water tanks (Figs. 5.19 and 5.20) are considered, the volume of paraffin containers to be placed should be adjusted carefully. Although the increasing PCM amount inside the tank increases the heat storage capacity, it reduces the volume of hot water output as the PCM containers take volume of the ready-to-use hot water.

As mentioned previously, the thermal stratification is one of the most important performance parameters for the thermally stratified heat storage tanks such as hot water tanks. Therefore, the effect of PCM placement inside the hot water on the thermal stratification should be investigated. The literature related to the use of PCM in the hot water tanks indicates that PCMs do not affect the degree of the thermal stratification of the tank. Table 5.6 shows the variation of Richardson and MIX numbers

Table 5.5 The Energy and Exergy Efficiencies for the Ordinary Tank and.

Experimental Conditions	Ordinary Tank		13 Capsules	
	η	ψ	η	ψ
$T_{m,g} = 80°C$, $V_{manto} = 2.5$ L/min, $V_{sebeke} = 5$ L/min	%87.36	%94.01	%22.03	%41.98
$T_{m,g} = 80°C$, $V_{manto} = 2.5$ L/min, $V_{sebeke} = 10$ L/min	%86.53	%93.57	%20.45	%39.16
$T_{m,g} = 80°C$, $V_{manto} = 2.5$ L/min, $V_{sebeke} = 15$ L/min	%85.89	%93.01	%18.34	%36.34
$T_{m,g} = 80°C$, $V_{manto} = 5$ L/min, $V_{sebeke} = 5$ L/min	%88.96	%94.56	%24.52	%44.17
$T_{m,g} = 80°C$, $V_{manto} = 5$ L/min, $V_{sebeke} = 10$ L/min	%87.78	%94.00	%22.13	%41.75
$T_{m,g} = 80°C$, $V_{manto} = 5$ L/min, $V_{sebeke} = 15$ L/min	%86.63	%93.14	%20.87	%38.34
$T_{m,g} = 80°C$, $V_{manto} = 7.5$ L/min, $V_{sebeke} = 5$ L/min	%89.34	%95.37	%27.19	%47.73
$T_{m,g} = 80°C$, $V_{manto} = 7.5$ L/min, $V_{sebeke} = 10$ L/min	%88.87	%94.72	%25.46	%44.79
$T_{m,g} = 80°C$, $V_{manto} = 7.5$ L/min, $V_{sebeke} = 15$ L/min	%88.14	%93.78	%23.08	%41.39

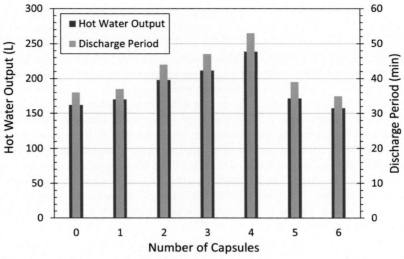

FIG. 5.20

The variation of hot water output with the number capsules for the case of $T_{mantle,in} = 80°C$.

Data from Ref. [8].

with the number of capsules placed inside the tank. It is clear from Table 5.6 that the degree of the thermal stratification does not change with the paraffin amount.

5.5.3 Closure

The PCMs placed inside the hot water tanks are one of the most effective methods to increase. The following key results are observed from the present case study:

- Although the temperature level inside the tank reduces with the increasing number of the paraffin capsules, the temperature distribution characteristics and the degree of the thermal stratification do not change significantly.
- For the vertical hot water storage tank, while the hot water output increases with the increasing number of paraffin capsules up to 13 paraffin capsules, it decreases after 13 paraffin capsules. This situation occurs as paraffin capsules decrease the volume of water inside the tank. After 13 paraffin capsules, although the stored energy amount inside the tank increases, the hot water output decreases because the stored energy in the paraffin cannot be transferred to the water passing through the capsules. 13 paraffin capsules increase the volume of obtained hot water from the tank by approximately 20% of the ordinary tank.
- For the horizontal hot water tank, the hot water output has increased up to four paraffin capsules. For the cases of five and six capsules, the hot water output has decreased substantially. Four paraffin capsules have increased the volume of hot water output by 40%.

Table 5.6 The Variation of the Richardson Number With the Number of Capsules.

Experimental Conditions	Number of Capsules						
	16	14	13	12	8	4	0
Richardson Number							
$T_{mantle,in} = 80°C$ $V_{mantle} = 2.5$ L/min	1.36	1.37	1.37	1.38	1.39	1.39	1.40
$T_{mantle,in} = 80°C$ $V_{mantle} = 5$ L/min	1.41	1.40	1.41	1.42	1.43	1.44	1.45
$T_{mantle,in} = 80°C$ $V_{mantle} = 7.5$ L/min	1.43	1.44	1.45	1.46	1.46	1.47	1.48
MIX Number							
$T_{mantle,in} = 80°C$ $V_{mantle} = 2.5$ L/min	0.45	0.44	0.45	0.46	0.46	0.46	0.47
$T_{mantle,in} = 80°C$ $V_{mantle} = 5$ L/min	0.46	0.47	0.47	0.48	0.48	0.48	0.49
$T_{mantle,in} = 80°C$ $V_{mantle} = 7.5$ L/min	0.48	0.48	0.49	0.49	0.49	0.49	0.50

5.6 Case Study 5: A Practical Application of the Encapsulated Ice Storage System

Ice storage systems (ice-SS), also named ice thermal energy storage, heat storage in the ice, and cold capacity storage in the ice, are generally integrated into the HVAC (heating, ventilating, and air-conditioning) system of the buildings to reduce the operating costs of the air-conditioning (AC) systems. They are also used for reducing the capacity of the chillers by shaving peak cooling loads. As mentioned in Section 2.8, the advantages of the ice-SSs are not only limited to reducing cooling costs and peak loads, but they also help to manage the grid supply and demand profiles and can be used as an emergency backup unit for cooling system. In this case study, an encapsulated ice-SS, which is integrated into a supermarket's AC system, is discussed in terms of thermodynamic and economic performance. The effect of different heat storage utilization strategies on the initial investment and operating costs is also discussed with the experimental observation from the supermarket facility.

5.6.1 System Description

This case study presents experimental observations of an encapsulated ice-SS that is integrated into the AC system used in the supermarket located in Ankara, Turkey. Fig. 5.21 shows the view of the supermarket. The supermarket facility has 15,000 m² of the closed area. 10,000 m² of the facility is cooled by the AC. The remaining 5000 m² is used for storage purposes and parking lots.

The working hour of the supermarket is 08:00—22:00. Fig. 5.22 demonstrates hourly peak cooling loads of the supermarket, and tariff rates are applied to the supermarket. Total daily peak load of the supermarket is 8332 kWh. The peak cooling load is 847 kW or 3049 MJ/h. The peak cooling demands are seen between 14:00 and 21:00. In the supermarket, there is a separate electricity subscription for the AC system. Thus, the effect of the ice-SS on the cooling costs is determined clearly.

FIG. 5.21

The view of the supermarket.

The image has been retrieved from Google Maps [11].

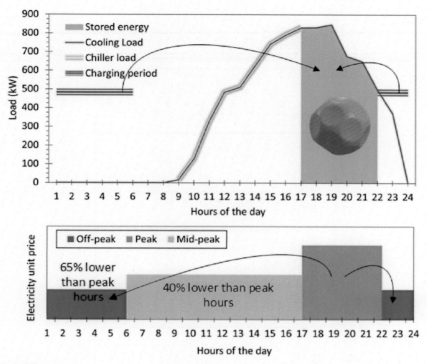

FIG. 5.22

The hourly peak cooling loads and tariff periods for the supermarket.

Data from Ref. [12].

In the electricity tariff used in ice storage—integrated AC system, the off-peak period is 65% lower than the peak load, which is 40% lower than the midpeak period. The difference between peak and off-peak periods indicates the potential of the ice-SS in reducing the cooling costs.

The meteorological data for the location of the supermarket are given in Table 5.7. When the meteorological conditions of the location of the supermarket and heat gains due to fresh air supply are considered, the supermarket is required to cool by using AC systems for 5 months in a year.

Fig. 5.23 shows the schematic view of the ice storage—integrated AC system. A 100 m³ ice storage tank is used in the system. There are 85,000 ice capsules, shown in Fig. 5.24, in the storage tank. The diameter of the capsule is 11 cm. An air gap is left inside the capsule to absorb the volume changes due to ice/water phase changing. The porosity of the tank is approximately 0.5. 35% ethylene glycol solution is used as the heat transfer fluid.

Table 5.7 Meteorological Data of Ankara for Between 1927 and 2018.

Months	Mean Temperature°C	Mean Highest Temperature°C	Highest Temperature°C	Mean Lowest Temperature°C	Lowest Temperature°C	Mean hours of Sunshine	Mean Rainy day
1	0.4	4.4	16.6	−3	−22.4	2.5	12.3
2	1.8	6.5	20.4	−2.2	−22.2	3.5	10.9
3	6.0	11.6	27.8	0.9	−19.2	5.2	11
4	11.4	17.3	31.1	5.6	−6.7	6.4	11.7
5	16	22.2	33.0	9.6	−1.6	8.4	12.6
6	20.1	26.6	37.0	13	3.8	10.2	8.8
7	23.5	30.2	41.0	15.9	4.5	11.3	3.8
8	23.3	30.3	40.4	16	6.3	11.6	2.7
9	18.7	26.0	36.0	11.8	2.5	9.2	3.9
10	12.9	19.8	33.3	7.2	−5.3	6.5	6.9
11	7.0	12.8	24.4	2.4	−13.4	4.4	8.5
12	2.6	6.6	20.4	−0.7	−18.0	2.3	11.6

Data from Ref. [13].

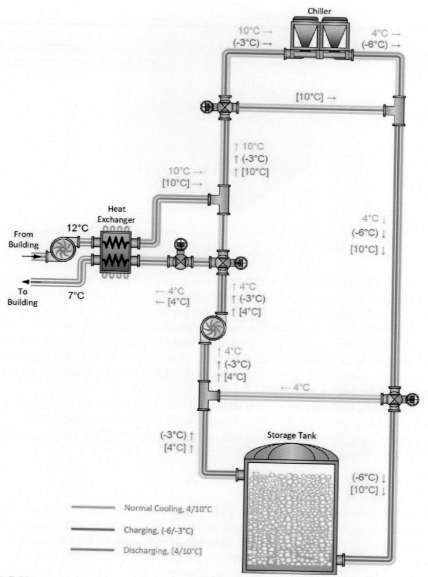

Chiller

10°C →
(-3°C) →

4°C →
(-6°C) →

[10°C] →

↑ 10°C
↑ (-3°C)
↑ [10°C]

10°C →
[10°C] →

Heat
Exchanger

From
Building

12°C

4°C ↓
(-6°C) ↓
[10°C] ↓

To
Building

7°C

← 4°C
← [4°C]

↑ 4°C
↑ (-3°C)
↑ [4°C]

↑ 4°C
↑ (-3°C)
↑ [4°C]

← 4°C

(-3°C) ↑
[4°C] ↑

Storage Tank

(-6°C) ↓
[10°C] ↓

Normal Cooling, 4/10°C

Charging, (-6/-3°C)

Discharging, [4/10°C]

FIG. 5.23

The schematic view of the ice storage—integrated AC system used in the supermarket.
AC, air-conditioning.

FIG. 5.24

The view of the ice capsules used in the ice storage system [14].

Two BLUEBOX TETRIS-W/LC 54.4 have been used in the system as a chiller unit. Chillers have the capability to work in dual mode by adjusting the evaporator temperature for working conditions, which are ice producing (charging) period and normal cooling (storing) period. During the normal cooling period, the HTF circulates at a temperature of 10/4°C. The cooling capacity of the chiller is 397.5 kW, and the power of the compressor is 125 kW. In the charging period, The HTF circulates at the temperature of −3/−6°C. The power of the chiller is 101 kW, and the cooling capacity of the chiller is 336 kW.

The cooling loads between 17:00 and 22:00, which are in the electricity peak periods, are stored by the ice-SS. The chillers are not operated when the electricity is expensive during the day. 3875 kWh of cooling capacity, which corresponds to 47% of the total cooling load (8332 kWh), is stored.

5.6.2 Results and Discussion

The effect of the ice-SS on the COP is studied first. Traditional AC systems are operated when the outdoor temperature, solar radiation, and relative humidity are higher. Higher values of the three parameters substantially reduce the COP of the chiller. As the ice-SSs allow to run the chillers during the night hours, they are provided to operate the chillers with higher COP values because the outdoor temperature and relative humidity are lower and there is no solar radiation. On the other hand, ice-SSs tend to reduce the COPs, as a lower evaporator temperature is needed to produce ice. At the location of the hypermarket, during the summer, there is a 10°C temperature difference between day and night. While the average temperature of the working fluid is −4.5°C during the ice production, it is 7°C during the standard AC system. Namely, the difference in the working fluid temperature is 11.5°C for the modes with and without ice-SS. While the change in the outdoor temperature has tended to increase the COP, the change in the working fluid temperature has tended to decrease it. The COP of the chiller group has been calculated as 3.18 and 3.32 for standard cooling and ice production periods, respectively. As can be seen from the COPs, the ice-SS has increased the COP by 4.4%. This also means ice-SS has the potential to reduce electricity consumption by 4.4%. When designing an ice-SS, the working fluid temperature and outdoor temperature should be taken into consideration to increase the economic performance of the system.

The electricity consumptions and operating costs of the ice storage—integrated AC system are given in Table 5.8. As expected, electricity consumption has substantially increased in the summer seasons with respect to AC usage. AC system has been operated for almost 6 months from late April to early October to meet the cooling need of the hypermarket. In the remaining months, the AC system has not been operated, and the consumed electricity has been used for the equipment using for heating and fresh air. The highest energy consumption is observed to be in July and August due to increased AC utilization. When the four summer seasons investigated are considered, the total amount of electricity consumption has been measured as 816 kWh, 916 kWh, 1076 kWh, and 847 kWh, respectively. In the summer seasons, the average consumption for midpeak, peak, and off-peak periods are 40.47 kWh, 16.04 kWh, and 117.60 kWh, respectively. 67.54% of the total electricity consumption has occurred in the off-peak period. The average consumption of 16.04 kWh during peak periods is consumed by pumps and fans used during the energy discharge period.

As seen from Table 5.8, the highest cost among the electricity tariffs is seen in the service tariff as it covers entire electricity consumption in all tariffs. In July 2019, the service tariff has been canceled by changing the regulation in the electricity tariff; therefore, the last three values are zero. As expected, the electricity consumption costs have increased between April and September with the increasing AC use. Other months when AC is not operated, there are only base costs due to fans and pumps used for heating purposes. The electricity cost in the off-peak hours is 41%−66% higher than the peak hours and 17%−35% higher than midpeak hours. These values show the potential of the ice-SS in shifting the cooling cost to the off-peak periods. Most of the electricity consumption for cooling purposes has been met with the lowest electricity unit price in the off-peak period.

Fig. 5.25 demonstrates the comparison of the electricity consumption costs. The operating cost of the standard AC (without ice-SS) has been calculated by assuming the electricity consumed in the off-peak period as if in the peak hours. The operating costs of the standard AC have been calculated by considering both with and without the COP effect. It is observed from Fig. 5.25 that the ice storage has decreased the AC operating cost between 41% and 50% during the summer when the cooling load in the electricity peak tariff hours has been stored. The effect of COP on the

Table 5.8 The Electricity Consumptions and Operating Costs of Ice Storage–Integrated AC System.

	Period	Electricity Consumption (kWh)				Electricity Cost (TRY)					
		Midpeak	Peak	Off-Peak	Total	Midpeak	Peak	Off-Peak	Service	Total (No Tax)	Total (Tax Included)
1	July 2016	97.191	39.419	171.484	308.094	4257	3061	5253	9,330	21,902	27,048
2	August 2016	59.578	22.033	162.122	243.733	2619	1715	3161	5,541	13,036	16,100
3	September 2016	62.909	23.504	178.321	264.734	2766	1829	3477	6,018	14,090	17,401
4	October 2016	7.589	4.769	3.182	15.540	334	371	62	355	1122	1386
5	November 2016	10.947	5.260	4.101	20.308	481	409	80	462	1432	1769
6	December 2016	11.637	5.555	4.179	21.371	512	432	81	486	1511	1866
7	January 2017	12.310	5.751	4.319	22.380	524	440	78	509	1551	1915
8	February 2017	10.675	5.171	3.840	19.686	454	395	69	448	1366	1687
9	March 2017	11.023	5.197	4.665	20.885	469	397	84	475	1425	1760
10	April 2017	7.501	4.395	2.510	14.406	319	336	45	327	1028	1269
11	May 2017	6.733	4.348	0.975	12.056	286	332	18	276	912	1126
12	June 2017	30.171	12.418	91.243	133.832	1283	949	1649	3042	6923	8550
13	July 2017	63.047	21.814	217.863	302.724	2682	1667	3937	6,882	15,167	18,731
14	August 2017	56.664	4.623	169.873	231.160	2410	353	3070	5,255	11,088	13,694
15	September 2017	38.910	36.508	146.452	221.870	1655	2789	2646	5,044	12,135	14,986
16	October 2017	9.276	5.332	13.087	27.695	398	409	242	629	1678	2073
17	November 2017	10.439	4.321	1.574	16.334	448	332	29	394	1203	1486

18	December 2017	11.448	5.308	4.030	20.786	492	408	75	502	1476	1823
19	January 2018	12.750	5.967	4.357	23.074	582	487	86	557	1712	2115
20	February 2018	11.589	5.170	4.055	20.814	529	422	80	502	1534	1894
21	March 2018	10.520	5.132	4.082	19.734	483	421	81	476	1461	1805
22	April 2018	7.623	4.877	1.168	13.668	369	422	25	330	1146	1415
23	May 2018	7.848	5.292	6.652	19.792	382	460	142	478	1462	1805
24	June 2018	51.507	19.193	161.372	232.072	2505	1669	3445	5,602	13,221	16,328
25	July 2018	65.891	24.560	214.615	305.066	3205	2136	4582	7,364	17,286	21,349
26	August 2018	56.656	19.704	204.393	280.753	3245	1900	5371	6,667	17,184	21,222
27	September 2018	48.025	21.381	155.584	224.990	3306	2422	5213	5,336	16,277	20,102
28	October 2018	7.589	4.527	1.090	13.206	660	632	49	313	1654	2043
29	November 2018	9.657	4.469	2.436	16.562	840	629	111	439	2019	2494
30	December 2018	11.743	5.363	5.052	22.158	1032	755	231	587	2605	3217
31	January 2019	12.554	5.766	4.667	22.987	1054	789	195	614	2652	3275
32	February 2019	12.220	5.497	4.400	22.117	1026	752	184	572	2534	3130
33	March 2019	12.096	5.537	4.242	21.875	1015	758	177	564	2515	3106
34	April 2019	7.222	4.349	1.280	12.851	584	582	50	332	1547	1911
35	May 2019	6.723	4.539	0.996	12.258	543	607	38	316	1504	1857
36	June 2019	31.895	13.731	92.274	137.900	2574	1835	3558	3,857	11,824	14,603
37	July 2019	48.776	17.873	179.475	246.124	4613	2772	8331	6,915	22,631	27,949
38	August 2019	45.051	15.866	168.811	229.728	5985	3,074	14,246	0	23,305	28,782
39	September 2019	40.983	13.439	141.995	196.417	5445	2,604	11,983	0	20,031	24,739
40	October 2019	8.354	5.367	15.411	29.132	1110	1040	1301	0	3450	4261
Total										**278,601**	**344,072**

Data from Ref. [12].

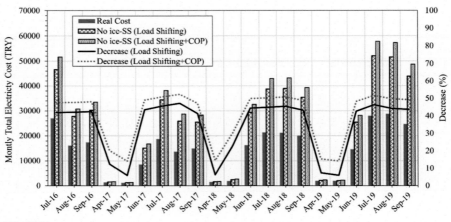

FIG. 5.25

The comparison of the electricity consumptions for the ice storage—integrated AC system and conventional AC. *AC*, air-conditioning.

Data from Ref. [12].

operating costs has been observed to be around 3.2%. In other words, the ice-SS has reduced the cooling cost by 3.2% thanks to decrease in outdoor temperature.

The ice storage—integrated AC system has consumed 344,072 TL (86,115 USD) of electricity for the observation period of 40 months. This is the actual operating costs of the ice storage—integrated AC system obtained from monthly electricity bills issued to the supermarket. For the case of without ice storage, the operating cost has been calculated as 1,167,680 TL. When the COP effect is excluded, the cost has been calculated to be 1,260,630 TL. The ice-SS has reduced the AC operating cost by 69.22% and 71.49% for the cases considered with and without ice-SS. Consequently, it is observed that the ice-SS has decreased the cooling cost by about 70% by storing cooling loads in the peak electricity tariff period.

The initial investment cost of the ice storage—integrated AC system is 376,715 USD (1.1 million TL). The initial cost of a conventional AC system is 202,320 USD (590,767 TL). In other words, the extra cost of ice storage integration is 174,395 USD (509,233 TL). This additional investment covers the storage tank, ice capsules, three-way valves, installment equipment, and the rise in the capacity of the pump. It is clear from the ice-SS's investment cost that an extra 86.22% investment has been required according to a normal AC system to store 46.30% of the total cooling load. A cost of 52.21 USD/kWh is required for storing a 1 kWh of cooling load. Fig. 5.26 shows the payback curve of the ice storage—integrated AC system. Values in the first 40 months are actual electricity costs obtained from issued electricity bills. Monthly costs in other months are projected by assuming a yearly 10% rise in the electricity prices. The ice-SS has paid its investment cost back in 27 months (in third summer season). At the end of 120 months

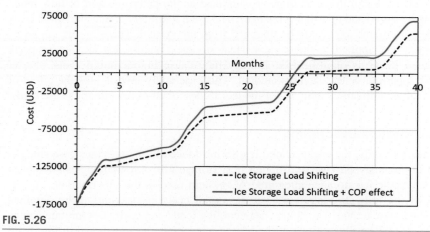

FIG. 5.26

The payback curve of the ice storage—integrated AC system. *AC,* air conditioning.

Data from Ref. [12].

(10 summer season), it is estimated that the ice storage—integrated AC system will spend 1.7 million TL of electricity. For the case of without the ice-SS, the electricity cost is calculated to be 5.4—5.9 million TL. Results show that the ice-SS has the potential to reduce electricity costs more than 4 million TL at the end of 10 years. When the initial cost is deducted from the saving supplied by ice-SS, the net profit is more than 3.22 million TL. When the 40-month experimental period has been taken into consideration, the AC operating cost of GIMSA hypermarket has been reduced more than 60% by storing the cooling loads in the peak electricity periods.

5.6.3 Closure

This case study presents the economic benefits of the ice-SS integrated into the AC system used in the supermarket. The following are the key findings of the current case study:

- The ice storage has achieved to shift the 67.54% of total consumed electricity for AC purposes to the off-peak periods.
- The differences in environmental (day and night) and working fluid temperatures due to the ice-SS have a great impact on the COP of the chillers. The COP of the chiller has increased by 4.4% thank to the ice-SS.
- The ice-SS has reduced the operating costs approximately by 60%.
- While the ice storage—integrated AC system has required 1.1 million TRY (1 USD = 2.92 TRY) of initial cost, a conventional AC system has required 590,767 TRY. 509,233 TRY of extra investment cost of the ice-SS has been paid back in 27 months (three summer seasons).
- At the end of 10 years, the ice-SS can achieve more than 3.22 million TL of saving.

5.7 Case Study 6: Effects of Heat Storage Utilization Strategy

The heat storage utilization strategy is defined as the determination of how much heat to be stored and how the stored heat to be used. It directly affects the initial costs of the system and savings to be obtained from the system. Therefore, it should be analyzed and applied to the system. Heat storage systems that are not properly designed in terms of utilization strategy may cause higher initial costs, shorter payback period, or less savings amount. This case study investigates the effect of heat storage utilization strategy for the ice-SS described in the previous case study.

5.7.1 System Description

In this case study, the encapsulated ice-SS introduced in the previous case study is used in the analysis. Twelve different heat storage utilization strategies have been applied to the ice storage—integrated AC system used in the supermarket. The effects of the utilization strategies on the initial costs and savings to be obtained have been researched. The hourly peak cooling loads and tariff structure are shown in Fig. 5.22. The illustrations of the heat storage strategies are shown in Fig. 5.27. Although these strategies seem specific for the supermarket studied, their aim and load distribution characteristics are the same for all heat storage systems. The following strategies are applied to the ice-SS:

- Full storage strategy (Fig. 5.27a): In the full storage strategy, the cooling load of the building is stored by the ice-SS. The primary purpose of this strategy is to use off-peak electricity. The chiller does not operate during the peak and midpeak periods. Full storage strategy achieves the highest savings. On the other hand, it requires higher capacity chillers and a higher volume storage tank, which both cause the higher initial costs. The significant issue in applying the full storage strategy is the requirement of the high-volume storage tank. Therefore, it is usually considered for the new buildings. It should be included in the planning stage of the building.
- Load leveling strategy (Fig. 5.27b): The main purpose of the load leveling strategy is to reduce the capacity of the chillers. A portion of the total cooling load of the building is stored to provide the load leveling. In this strategy, the chiller is operated perpetually during the day at a constant load. This strategy is effective to use when the grid supply power is not enough. It minimizes the power requirements of the chillers and, of course, their capacities. Therefore, it has considerable potential to reduce the initial cost of the chillers. It also reduces the auxiliary equipment and systems used in the buildings, such as power generators, power cables, and so on.
- Partial storage for peak periods (Fig. 5.27c): This strategy aims to avoid the peak electricity periods for reducing cooling costs and electricity peak loads. The chillers are not operated during the electricity peak tariff periods. It is a specific type of partial storage strategy. This strategy provides maximum benefit when both peak cooling loads and electricity peak tariff periods match.
- Partial storage strategy (Fig. 5.27d): In the partial storage strategy, a portion of the cooling load of building is stored with the ice-SS. This strategy is used for shifting a certain portion of the cooling loads to off-peak periods. It is useful when there is a limited space for the storage tank. The stored cold capacity is generally used for the precooling of heat transfer fluid before it enters the chiller.

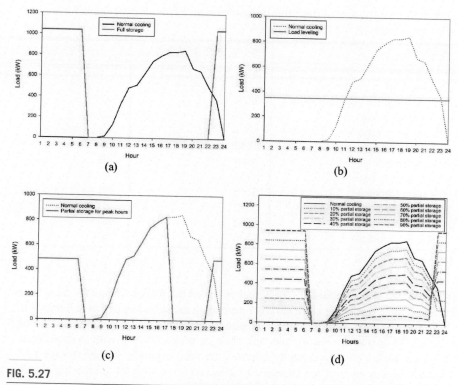

FIG. 5.27

The illustration of the heat storage utilization strategies for the ice storage system (a) full storage, (b) load leveling, (c) partial storage for peak hours, and (d) partial storage.

Data from Ref. [15].

The amounts of the stored energy and their percentages are given in Table 5.9. To store the cooling loads in the electricity peak tariff periods, 47% of the total cooling storage is stored. For the load leveling, 33% of the total cooling load should be stored. As can be seen from these values, although the partial storage for peak periods and load leveling strategies are the partial storage, they use the stored energy differently. Therefore, their impacts on the initial costs and savings are different.

5.7.2 Initial Investment Cost for the Ice Storage System

As mentioned in the previous case study, the partial storage for peak hours strategy has been applied to the ice-SS in the supermarket. The initial investment cost of the ice storage—integrated AC system is 35,000 USD. The initial cost of a conventional AC system is 195,000 USD. In other words, the extra cost of ice storage integration is

Table 5.9 The Amount of the Storage Energies and Their Percentages for the Strategies Given in Fig. 5.27

Strategy	Stored Energy (kWh)	Percentage (%)
Full storage	8332	100
Load leveling	2777	33
Partial storage for the peak hours	3876	47
10% partial storage	1168	10
20% partial storage	1964	20
30% partial storage	2760	30
40% partial storage	3556	40
50% partial storage	4352	50
60% partial storage	5148	60
70% partial storage	5943	70
80% partial storage	6739	80
90% partial storage	7535	90

Modified from Ref. [15].

160,000 USD. These values are the actual values of the partial storage for peak hours strategy. Other values are generated by referencing the actual values as in Table 5.10.

Table 5.10 gives the cost of the main equipment in the ice storage—integrated AC system. The cost of the ice storage tank changes proportionally with storage capacity, as the higher storage capacity requires the proportionally higher storage tank. Similarly, the number of ice storage capsules increases with the increasing storage capacity. Therefore, the cost of capsules increases proportionally with the storage capacity. Due to the increasing number of capsules and tank volume, pressure drops

Table 5.10 The Costs of the Equipment Used in the Ice Storage—Integrated Air-Conditioning System and Their Changes With the Heat Storage Utilization Strategy.

Equipment	Cost (USD)	Cost Changing With the Strategy
Ice storage tank	75,000	Proportional to storage capacity
Capsules	35,000	Proportional to storage capacity
Chiller	150,000	Proportional to storage capacity
Pump	10,000	Proportional to storage capacity
Heat exchanger	20,000	Constant
Heat transfer fluid	15,000	Proportional to storage capacity
Installation and other equipment	50,000	Constant

Modified from Ref. [15].

in the system increase; this increase requires a higher capacity pump. So the cost of the pump increases proportionally with storage capacity. The cost and capacity of the heat exchanger used for cooling the heat transfer fluid circulated in the building are constant as its capacity depends on the building energy demand. The cost of the chillers is directly related to the cooling peak load. Therefore, the chiller costs change proportionally with the peak cooling capacity.

5.7.3 Results and Discussion

This case study presents the effect of heat storage utilization strategy on the initial and energy consumption costs for the ice-SS integrated into the supermarket's AC system. The initial cost of the ice-SS, payback period, and total savings amount at the end of 10 years have been calculated through the peak cooling load of the building. The following assumptions have been applied to the analysis:

- The analysis has been done through the hourly peak loads of the building.
- Year rise in the electricity unit prices has been assumed to be 5%.
- The annual service cost is 10,000 USD and yearly rise is 5% for it.
- Unexpected and unplanned troubles and stops are not taken into consideration.
- Taxes are not included into calculations.

Initial costs of the ice storage—integrated AC system to be used in the building and its extra investment costs for the ice-SS are given in Table 5.11. Fig. 5.28 shows the initial costs of the main element of the system. While the lowest chiller cost and

Table 5.11 The Costs of the Equipment Used in the Ice Storage—Integrated Air-Conditioning System and Their Changes With the Heat Storage Utilization Strategy.

Strategy	Initial Cost (USD)	Extra Investment for the Ice Storage (USD)
Full storage	628,976	433,976
Load leveling	209,585	14,858
Partial storage for the peak hours	355,000	160,000
10% partial storage	199,485	4485
20% partial storage	226,403	31,403
30% partial storage	253,140	58,140
40% partial storage	279,877	84,877
50% partial storage	328,472	133,482
60% partial storage	388,653	193,653
70% partial storage	448,644	253,644
80% partial storage	508,634	313,634
90% partial storage	569,805	373,805

MBE: Mass balance equation; EBE: Energy balance equation; EnBE: Entropy balance equation; ExBE: Exergy balance equation.
Modified from Ref. [15].

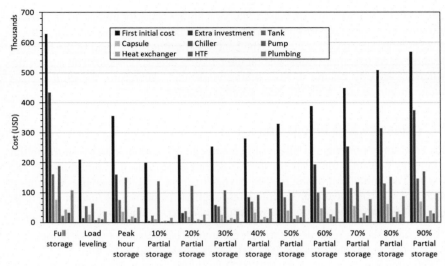

FIG. 5.28

The initial costs of the ice storage—integrated AC systems and system's equipment. *AC,* air-conditioning.

Data from Ref. [15].

capacity are seen in the load leveling strategy, the highest chiller cost and capacity are seen in the full storage.

Fig. 5.29 shows the electricity costs for the conventional AC system and the ice storage—integrated AC system. The cooling costs decrease with the increasing storage capacity. In other words, economic savings in the cooling costs rise with the increasing storage capacity. Load leveling and partial storage for peak hours strategies provide more savings than the standard partial storage strategies because they use the stored energy in the peak tariff hours more than the standard partial storage strategies, which have the same cooling capacity. The partial storage for peak periods strategy has provided about 23% more economic savings than the 50% standard partial storage. The load leveling strategy has provided approximately 30% more savings than the 30% standard partial storage. Fig. 5.30 illustrates the variation of the daily cooling cost and unit cooling cost with the storage capacity. The impacts of the partial storage for peak periods and load leveling are clearly seen in Fig. 5.30. Consequently, if the primary purpose of the heat storage is to reduce the energy consumption costs, the stored energy should be used in the electricity peak periods.

The payback period and the savings amount for the different heat storage utilization strategy are shown in Fig. 5.31. While the highest savings is seen in full storage strategy with the value of 1.4 million USD, the lowest savings is in the 10%

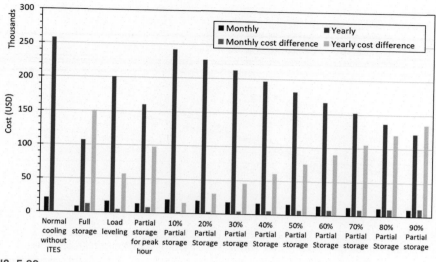

FIG. 5.29

The comparison of the monthly and yearly electricity consumption costs for the conventional AC and the ice storage—integrated AC system. *AC*, air-conditioning.

Data from Ref. [15].

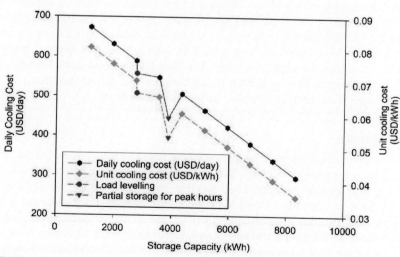

FIG. 5.30

The variation of the daily cooling cost and unit cooling cost with the storage capacity.

Data from Ref. [15].

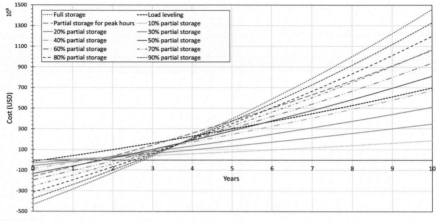

FIG. 5.31

The variation of the daily cooling cost and unit cooling cost with the storage capacity.

Data from Ref. [15].

standard partial storage strategy. As expected, the savings increases with the increasing storage capacity. Also, the load leveling and partial storage for the peak hour strategies provides more savings than the standard partial strategies that have the same capacity with them. When the heat storage utilization strategies are evaluated through payback periods, the shortest payback period is seen in the 10% partial storage. The payback period extends with the increasing storage capacity. Therefore, the longest payback period is seen in the full storage strategy, although it provides the highest savings. The load leveling and partial storage for peak period strategies have paid back the investment cost earlier than the standard partial storage strategies that have the same capacity as them.

5.7.4 Closure

This situation extends the payback period of the heat storage system. On the other hand, in the long term (a decade), the amount of savings has increased with the increasing heat storage capacity. Load leveling and partial storage for peak hours strategies have indicated that besides the amount of energy stored, how the stored energy is used also has a significant effect on the economic performance of the ice-SS because load leveling and partial storage for peak hours strategies have provided a lower payback period and a higher profit at the end of 10 years than ordinary partial storage strategies at the same storage capacity. The system works with the highest economic performance when the peak cooling demand hours and electricity peak tariff hours match. While the shortest payback period is seen in both load leveling and 10% partial storage strategies with about 1.5 years, the longest payback period is seen in full storage strategy with about 3.1 years.

5.8 Case Study 7: Thermodynamic Analysis of an Ice Storage System

As explained in Chapter 2 and previous case studies, ice-SSs, which are one of the most used heat storage applications in the buildings, provide significant benefits to reduce energy consumption costs and shift the peak loads to off-peak periods. Additionally, the thermodynamic analysis of the system is fundamental to the system in terms of system design and economic performance. Thermodynamically improved heat storage systems provide a lower initial cost of equipment and more savings. Also, the COP of the system is very critical for the economic performance of the system. Therefore, it should be calculated and optimized carefully for a better system. This case study presents a detailed thermodynamic analysis for ice-SSs. Also, parametric studies are performed to investigate the effects of the design and operating conditions on the system performance.

5.8.1 System Description

The schematic diagrams of the working periods for an ice-SS are shown in Fig. 5.32. The system works under three modes: normal cooling (storing), charging, and discharging periods. In the normal cooling period, also called storing period (Fig. 5.32a), the building is cooled by the chillers as if the standard AC system. The cooled heat transfer fluid is circulated between the chiller and heat exchanger. Cooling demand of the building is met directly by air conditioners. In the charging period (Fig. 5.32b), the chillers cool down the heat transfer fluid at the temperature below 0°C, and the heat transfer fluid is pumped to the storage tank. Thus, ice is produced. In the discharging period (Fig. 5.32c), the stored ice is used for cooling down the heat transfer fluid that is circulated between storage tank and heat exchanger. The cooling demand of the building is met by stored ice.

When the environmental and operating conditions of the ice storage—integrated AC system are considered, the following two points should be carefully examined. For a standard AC system and normal cooling period of the ice storage—integrated AC system, the chillers work during the daytime when the outdoor temperature, relative humidity, and solar radiation are higher. Higher values of those cause lower COP values. During the charging period, the chillers operate during the nighttime when outdoor temperature and relative humidity are lower than daytime. Lower values of those provide a higher COP for chiller. On the other hand, to solidify the water, a lower evaporator is required, which causes a lower COP for chiller. Eventually, for ice-SSs and other cold heat storage systems, the operating conditions should be adjusted by considering environmental conditions. The lower COP to be obtained provides a lower electricity consumption.

5.8.2 Details of the Building

In this case study, an office building in Istanbul, Turkey is studied. The office building has a 17,670 m² closed area with an aspect ratio of 1.36. The building has 12

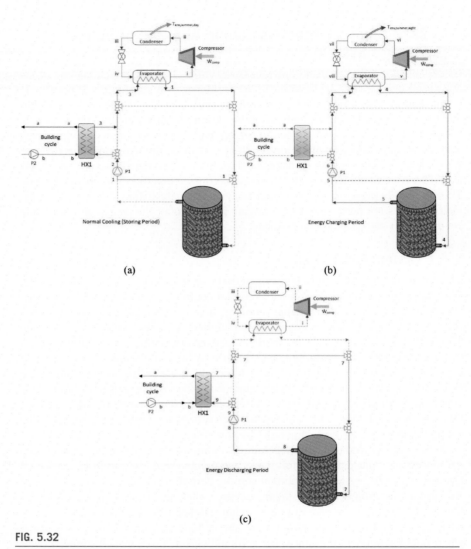

FIG. 5.32

The schematic view of the ice storage system for the operating periods: (a) charging, (b) discharging, and (c) normal cooling.

floors plus four underground parking levels. The building is open 12 h per day from 08:00–20:00 and 300 days per year. The parking lots are not air-conditioned. The hourly peak cooling loads of the building are shown in Fig. 5.33. It is assumed that all cooling load of the buildings is stored by the ice-SS. In other words, the full storage strategy has been applied to the system. The chiller only works at off-peak hours. The cooling demand of the building is met by the stored heat. The total daily cooling load is 13,636 kWh, and the peak cooling load is 1465 kW.

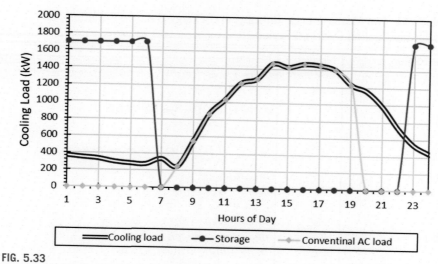

FIG. 5.33

The hourly peak cooling loads of the building.

Data from Refs. [16,17].

5.8.3 Thermodynamic Analysis

Thermodynamic analysis, based on energy and exergy calculations, is a significant tool for assessing heat storage systems. The COP is the most significant performance criterion for the cooling and heat pump systems. Therefore, the energetic and exergetic COPs should be calculated for the heat storage—integrated AC system to determine the effect of heat storage integration. Here, a complete thermodynamic analysis for the ice storage—integrated AC system is applied the ice-SS. Unlike the previous case studies, the ice-SS is integrated into the HVAC of an office building. The state numbers presented in Fig. 5.32 are used in the thermodynamic analysis. The thermodynamic balance equations for each element of the system are given in the following Tables 5.12—5.14.

The main difference in these balance equations is the environmental conditions. For the normal cooling period, the environmental temperature (T_0) is the daytime outdoor temperature. During the charging period, the environmental temperature (T_0) is the nighttime outdoor temperature. The refrigeration cycles have been modeled separately for normal cooling and charging period. Thus, the power demands of the compressor have been calculated for the standard AC and ice production mode. The difference in the compressor power mainly causes the difference in the COP.

The COP of the ice storage—integrated AC system should be defined by considering total cold outputs and total energy inputs in the daily time frame since the energy inputs and outputs occur in different periods. The energetic and exergetic COPs of the system for the working periods are defined in Table 5.15. The equations in Table 5.15 define the momentary COP values of the system for the operating periods under steady-state conditions.

Table 5.12 The Mass, Energy, Entropy, and Exergy Balance Equations for the Normal Cooling (Storing) Period.

Evaporator	MBE	$\dot{m}_3 = \dot{m}_1$
	EBE	$\dot{m}_3 h_3 + \dot{Q}_{eva} = \dot{m}_1 h_1$
	EnBE	$\dot{m}_3 s_3 + \left(\frac{\dot{Q}_{eva}}{T_{eva}}\right) + \dot{S}_{gen,eva} = \dot{m}_1 s_1$
	ExBE	$\dot{m}_3 ex_3 + \left(\frac{T_0}{T_{eva}} - 1\right)\dot{Q}_{eva} = \dot{m}_1 ex_1 + \dot{E}x_{dest,eva}$
Pump	MBE	$\dot{m}_1 = \dot{m}_2$
	EBE	$\dot{m}_1 h_1 + \dot{W}_{P1} = \dot{m}_2 h_2$
	EnBE	$\dot{m}_1 s_1 + \dot{S}_{sen,P1} = \dot{m}_2 s_2$
	ExBE	$\dot{m}_1 ex_1 + \dot{W}_P = \dot{m}_2 ex_2 + \dot{E}x_{dest,P1}$
HX1	MBE	$\dot{m}_2 = \dot{m}_3 \quad and \quad \dot{m}_a = \dot{m}_b$
	EBE	$\dot{m}_2 h_2 + \dot{m}_a h_a = \dot{m}_3 h_3 + \dot{m}_b h_b$
	EnBE	$\dot{m}_2 s_2 + \dot{m}_a s_a + \dot{S}_{gen,HX1} = \dot{m}_3 s_3 + \dot{m}_b s_b$
	ExBE	$\dot{m}_2 s_2 + \dot{m}_a s_a = \dot{m}_3 s_3 + \dot{m}_b s_b + Ex_{dest,HX1}$
Compressor	MBE	$\dot{m}_i = \dot{m}_{ii}$
	EBE	$\dot{m}_i h_i + \dot{W}_{komp} = \dot{m}_{ii} h_{ii}$
	EnBE	$\dot{m}_i s_i + \dot{S}_{gen,comp} = \dot{m}_{ii} s_{ii}$
	ExBE	$\dot{m}_i ex_i + \dot{W}_{komp} = \dot{m}_{ii} ex_{ii} + \dot{E}x_{dest,comp}$
Condenser	MBE	$\dot{m}_{ii} = \dot{m}_{iii}$
	EBE	$\dot{m}_{ii} h_{ii} = \dot{m}_{iii} h_{iii} + \dot{Q}_{con}$
	EnBE	$\dot{m}_{ii} s_{ii} + \dot{S}_{gen,con} = \dot{m}_{iii} s_{iii} + \left(\frac{\dot{Q}_{con}}{T_{con}}\right)$
	ExBE	$\dot{m}_{ii} ex_{ii} = \dot{m}_{iii} ex_{iii} + \left(1 - \frac{T_0}{T_{con}}\right)\dot{Q}_{con} + \dot{E}x_{dest,con}$
Expansion valve	MBE	$\dot{m}_{iii} = \dot{m}_{iv}$
	EBE	$\dot{m}_{iii} h_{iii} = \dot{m}_{iv} h_{iv}$
	EnBE	$\dot{m}_{iii} s_{iii} + \dot{S}_{gen,ev} = \dot{m}_{iv} s_{iv}$
	ExBE	$\dot{m}_{iii} ex_{iii} = \dot{m}_{iv} ex_{iv} + \dot{E}x_{dest,ev}$

MBE: Mass balance equation; EBE: Energy balance equation; EnBE: Entropy balance equation; ExBE: Exergy balance equation.

The overall COP values for the ice storage–integrated AC system can be written as

$$COP_{ove} = \frac{\left\{\sum_{t=0}^{t=t_{nc}}\left[\dot{m}_a(h_b - h_a)t\right]\right\} + \left\{\sum_{t=0}^{t=t_{disch}}\left[\dot{m}_a(h_b - h_a)t\right]\right\}}{\left\{\sum_{t=0}^{t=t_{nc}}\left(\dot{W}_{comp,nc} + \dot{W}_{P1} + \dot{W}_{P2}\right)t\right\} + \left\{\sum_{t=0}^{t=t_{ch}}\left(\dot{W}_{comp,ch} + \dot{W}_{P1}\right)\right\} + \left\{\sum_{t=0}^{t=t_{disch}}\left(\dot{W}_{P1} + \dot{W}_{P2}\right)t\right\}}$$

(5.29)

Table 5.13 The Mass, Energy, Entropy, and Exergy Balance Equations for the Charging Period.

Evaporator	MBE	$\dot{m}_6 = \dot{m}_4$
	EBE	$\dot{m}_6 h_6 + \dot{Q}_{eva} = \dot{m}_4 h_4$
	EnBE	$\dot{m}_6 s_6 + \left(\dfrac{\dot{Q}_{eva}}{T_{eva}}\right) + \dot{S}_{gen,eva} = \dot{m}_4 s_4$
	ExBE	$\dot{m}_6 ex_6 + \left(\dfrac{T_0}{T_{eva}} - 1\right)\dot{Q}_{eva} = \dot{m}_4 ex_4 + \dot{Ex}_{dest,eva}$
Storage tank	MBE	$\dot{m}_4 = \dot{m}_5$
	EBE	$\dot{m}_4 h_4 + \dot{Q}_{ST} = \dot{m}_5 h_5$
	EnBE	$\dot{m}_4 s_4 + \left(\dfrac{\dot{Q}_{ST}}{T_0}\right) + \dot{S}_{gen,ST} = \dot{m}_5 h_5$
	ExBE	$\dot{m}_4 ex_4 + \left[1 - \left(\dfrac{T_0}{273}\right)\right]\dot{Q}_{ST} = \dot{m}_5 ex_5 + \dot{Ex}_{dest,ST}$
Pump	MBE	$\dot{m}_5 = \dot{m}_6$
	EBE	$\dot{m}_5 h_5 + \dot{W}_{P1} = \dot{m}_6 h_6$
	EnBE	$\dot{m}_5 s_5 + \dot{S}_{gen,P1} = \dot{m}_6 s_6$
	ExBE	$\dot{m}_5 ex_5 + \dot{W}_{P1} = \dot{m}_6 ex_6 + \dot{Ex}_{dest,P1}$
Compressor	MBE	$\dot{m}_v = \dot{m}_{vi}$
	EBE	$\dot{m}_v h_v + \dot{W}_{comp} = \dot{m}_{vi} h_{vi}$
	EnBE	$\dot{m}_v s_v + \dot{S}_{gen,comp} = \dot{m}_{vi} s_{vi}$
	ExBE	$\dot{m}_v ex_v + \dot{W}_{comp} = \dot{m}_{vi} ex_{vi} + \dot{Ex}_{dest,comp}$
Condenser	MBE	$\dot{m}_{vi} = \dot{m}_{vii}$
	EBE	$\dot{m}_{vi} h_{vi} = \dot{m}_{vii} h_{vii} + \dot{Q}_{con}$
	EnBE	$\dot{m}_{vi} s_{vi} + \dot{S}_{gen,comp} = \dot{m}_{vii} s_{vii} + \left(\dfrac{\dot{Q}_C}{T_C}\right)$
	ExBE	$\dot{m}_v ex_v + \dot{W}_{comp} = \dot{m}_{vi} ex_{vi} + \dot{Ex}_{dest,comp}$
Expansion valve	MBE	$\dot{m}_{vii} = \dot{m}_{viii}$
	EBE	$\dot{m}_{vii} h_{vii} = \dot{m}_{viii} h_{viii}$
	EnBE	$\dot{m}_{vii} s_{vii} + \dot{S}_{gen,ev} = \dot{m}_{viii} s_{viii}$
	ExBE	$\dot{m}_{vii} ex_{vii} = \dot{m}_{viii} ex_{viii} + \dot{Ex}_{dest,ev}$

MBE: Mass balance equation; EBE: Energy balance equation; EnBE: Entropy balance equation; ExBE: Exergy balance equation.

Here, the numerator in Eq. (5.29) is the total cold output. The term on the left is the cold output during the normal cooling period (storing period). The term on the right is the cold output during the discharging period. The denominator in Eq. (5.29) is the total energy inputs in a daily cycle. The terms on the left are the energy input during

Table 5.14 The Mass, Energy, Entropy, and Exergy Balance Equations for the Discharging Period.

Storage Tank	MBE	$\dot{m}_7 = \dot{m}_8$
	EBE	$\dot{m}_7 h_7 = \dot{m}_8 h_8 + \dot{Q}_{ST}$
	EnBE	
		$\dot{m}_7 s_7 + \dot{S}_{gen,ST} = \dot{m}_8 s_8 + \left(\frac{\dot{Q}_{ST}}{T_0}\right)$
	ExBE	$\dot{m}_7 ex_7 + \left[1 - \left(\frac{T_0}{273}\right)\right]\dot{Q}_{ST} = \dot{m}_8 ex_8 + \dot{Ex}_{dest,ST}$
Pump	MBE	$\dot{m}_8 = \dot{m}_8$
	EBE	$\dot{m}_8 h_8 + \dot{W}_{P1} = \dot{m}_9 h_9$
	EnBE	$\dot{m}_8 s_8 + \dot{S}_{gen,P1} = \dot{m}_9 s_9$
	ExBE	$\dot{m}_8 ex_8 + \dot{W}_{P1} = \dot{m}_9 ex_9 + \dot{Ex}_{dest,P1}$
HX1	MBE	$\dot{m}_9 = \dot{m}_7 \quad and \quad \dot{m}_a = \dot{m}_b$
	EBE	$\dot{m}_9 h_9 + \dot{m}_a h_a = \dot{m}_7 h_7 + \dot{m}_b h_b$
	EnBE	$\dot{m}_9 s_9 + \dot{m}_a s_a + \dot{S}_{gen,HX1} = \dot{m}_7 s_7 + \dot{m}_b s_b$
	ExBE	$\dot{m}_9 s_9 + \dot{m}_a s_a = \dot{m}_7 s_7 + \dot{m}_b s_b + Ex_{dest,HX1}$

MBE: Mass balance equation; EBE: Energy balance equation; EnBE: Entropy balance equation; ExBE: Exergy balance equation.

Table 5.15 Energetic and Exergetic COPs for Normal Cooling, Charging, and Discharging Periods.

Energetic COP	Exergetic COP
$COP_{nc} = \frac{\dot{m}_a(h_b - h_a)}{\dot{W}_{comp,nc} + \dot{W}_{P1} + \dot{W}_{P2}}$	$COP^{ex}_{nc} = \frac{\dot{m}_a(ex_b - ex_a)}{\dot{W}_{comp,nc} + \dot{W}_{P1} + \dot{W}_{P2}}$
$COP_{ch} = \frac{\dot{m}_4(h_5 - h_4)}{\dot{W}_{comp,ch} + \dot{W}_{P1}}$	$COP^{ex}_{ch} = \frac{\dot{m}_4(ex_5 - ex_4)}{\dot{W}_{comp,ch} + \dot{W}_{P1}}$
$COP_{disch} = \frac{\dot{m}_a(h_b - h_a)}{\dot{m}_7(h_7 - h_8) + \dot{W}_{P1} + \dot{W}_{P2}}$	$COP^{ex}_{disch} = \frac{\dot{m}_a(ex_b - ex_a)}{\dot{W}_{P1} + \dot{W}_{P2}}$

the normal cooling period. The terms on the middle are the energy input during the charging period, and the last terms on the right are the energy input during the discharging period. It is clear from Eq. (5.29) that the COP and efficiency values for the energy storage systems should be performed by considering time periods to be applied to the system. The overall exergetic COP of the system is defined as

$$COP^{ex}_{ove} = \frac{\left\{\sum_{t=0}^{t=t_{nc}}\left[\dot{m}_a(ex_b - ex_a)t\right]\right\} + \left\{\sum_{t=0}^{t=t_{disch}}\left[\dot{m}_a(ex_b - ex_a)t\right]\right\}}{\left\{\sum_{t=0}^{t=t_{nc}}\left(\dot{W}_{comp,nc} + \dot{W}_{P1} + \dot{W}_{P2}\right)t\right\} + \left\{\sum_{t=0}^{t=t_{ch}}\left(\dot{W}_{comp,ch} + \dot{W}_{P1}\right)\right\} + \left\{\sum_{t=0}^{t=t_{disch}}\left(\dot{W}_{P1} + \dot{W}_{P2}\right)t\right\}}$$

(5.30)

The following assumptions have been applied to the thermodynamic analysis:

- The reference conditions are $T_0 = 25°C$ and $P_0 = 101.325$ kPa.
- The pressure drops in the system elements and pipes have been neglected.
- The system works under steady-state conditions.
- The thermophysical properties of the working fluids are temperature dependent.

5.8.4 Results and Discussion

The state point tables for normal cooling, charging, and discharging periods are given in Table 5.16. When these values are written into the balance equations, the exergy destructions and entropy generations can be obtained. Also, since there is

Table 5.16 The State Point Table for the System.

State Point	Working Fluid	T (K)	P (kPa)	h (kJ/kg)	s (kJ/kg K)	ex (kJ/kg)
1	Ethylene glycol	277	105	1058	–	6.66
2	Ethylene glycol	277	106.5	1058	–	6.66
3	Ethylene glycol	283	105	1082	–	4.26
4	Ethylene glycol	267	105	1018	–	4.91
5	Ethylene glycol	270	102	1030	–	3.82
6	Ethylene glycol	270	105	1030	–	3.81
7	Ethylene glycol	283	106.5	1082	–	4.26
8	Ethylene glycol	277	102	1058	–	6.66
9	Ethylene glycol	277.1	106.5	1058	–	6.62
13	Ethylene glycol	267	105	1018	–	0.60
14	Ethylene glycol	270	102	1030	–	0.27
15	Ethylene glycol	270	105	1030	–	0.26
i	R410a	266	196.8	431.6	1.99	–
ii	R410a	329.4	899.8	475.7	1.98	–
iii	R410a	306	887.4	453.2	1.91	–
iv	R410a	265	200.7	430.7	1.99	–
v	R410a	259	130.1	427.7	2.03	–
vi	R410a	317.8	581	469.4	2.01	–
vii	R410a	291	572.1	445	1.93	–
viii	R410a	258	132.9	426.9	2.02	–
a	Water	285	110	3044	0.18	−9.7
b	Water	280	110	3034	0.11	−14.17

Data from Ref. [17].

no database for the specific entropy values of the ethylene glycol on the EES, the entropy generation values have been calculated by using the relation between the exergy destruction and entropy generation (Eq. 5.32). The specific exergy for the ethylene glycol has been calculated by the following equation. The powers of the pump 1 (P1) and pump 2 (P2) are 15 and 10 kW, respectively.

$$ex = c_p \left[(T - T_0) - T_0 \ln \left(\frac{T}{T_{env}} \right) \right] \tag{5.31}$$

$$\dot{Ex}_{dest} = T_0 \dot{S}_{gen} \tag{5.32}$$

The COP values for the standard AC and ice storage—integrated AC system are given in Table 5.17. When comparing the COP values for the standard AC and ice storage—integrated AC system, the advantage of the ice storage is seen. While the COP of the standard AC is 2.97, the COP of the ice storage—integrated AC system is 3.50, which is 15% higher. In other words, when all cooling loads of the building are stored by ice storage, the cooling is met by almost 15% lower electricity consumption along with the economic savings due to electricity tariff. The overall COP for the ice storage—integrated AC is calculated to be 3.14, which is approximately 6% higher than the standard AC. Namely, when the energy demands of the pumps are included in the calculation, the ice-SS provides lower energy consumption. During the discharging period, since there is no compressor work and refrigeration cycle, we cannot consider the COP. Therefore, we are using efficiency during the discharging time. While the energy efficiency of the discharging period 96%, which is almost ideal, the exergy efficiency is 4%.

The environmental conditions have the biggest impact on the COP of the ice storage—integrated AC system. Therefore, the effects of daytime and nighttime outdoor temperatures on the COP have parametrically studied. Fig. 5.34 shows the variation of COPs with the outdoor temperatures. As seen from Fig. 5.34, when the temperature difference between night and day increases, the overall COP of the ice storage—integrated AC system increases. For the same nighttime outdoor temperature, the overall COP values increase with the decreasing daytime temperature.

Table 5.17 The Energetic and Exergetic COP Values for the Standard AC and Ice Storage—Integrated AC.

	Energetic	Exergetic
COP_{nc}	2.97	0.27
COP_{ch}	3.50	1.31
η_{disch}	0.94	0.04
COP_{ove}	3.14	0.85

AC, *air-conditioning.*

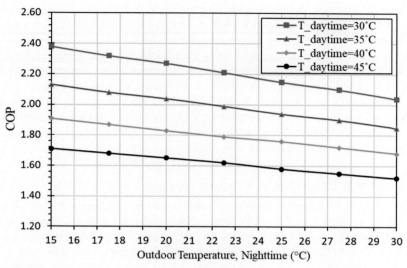

FIG. 5.34

The variation of the overall COP with daytime and nighttime outdoor temperatures.

5.8.5 Closure

Along with the economic savings and grid management capability of the ice-SSs, they can reduce the electricity consumption of the chiller due to changes in COP. The standard ACs are used during the daytime when the outdoor temperature and relative humidity are higher. These values substantially lower the COP of the chiller. Since ice-SSs are provided to work during the nighttime when the outdoor temperature and relative humidity are lower, they have the potential to increase the COP. A higher COP means a lower electricity consumption for refrigeration and heat pump systems. The observations that are obtained from the present case study are emphasized that the ice-SS integrated into the HVAC systems of an office building in Istanbul, Turkey, has increased the COP of the chiller by 15%. When all electricity consumptions are taken into consideration, the increase in the COP is 6%. Namely, while the ice-SS can cool down the building at a lower cost, it also has reduced electricity consumption. When the outdoor temperature difference between daytime and nighttime increases, the COP tends to increase. Ice-SSs can provide more benefits at locations where the difference in daytime and nighttime outdoor temperatures is higher.

5.9 Case Study: Numerical Analysis of Heat Storage Capsules

CFD and multiphysics analysis software is an essential part of the R&D and scientific research activities. Today, they are commonly used and integrated into R&D

and research activities, as they provide more accurate results to researchers and engineers related to systems and procedures. As in every field and sector, they are used in heat storage systems, too.

Encapsulation of the storage medium is a common technique in the heat storage systems. There are many types of the capsules, which are based on prismatic, cylindrical, and spherical geometries. Cylindrical and spherical ones are generally used due to their better heat transfer characteristics [18]. To enhance the heat transfer characteristics of the capsules, the geometrically modified capsules are commonly used. In this case study, the geometrically modified heat storage capsules have been investigated numerically.

5.9.1 System Description: Physical Domain

The capsule is at the center of the control volume, and the ethylene glycol flows over the capsule as shown in Fig. 5.35. The control volume has been selected from Ref. [19], which is $35 \times 35 \times 35$ cm of dimensions. The physical domain is modeled 3-D in full scale. The geometrically modified capsules given in Table 5.18 have been modeled. The volumes of the capsules are equal. Therefore, only the hydraulic diameter of the capsule has been changed between 107.05 mm (-2.68%) and 114.46 mm ($+4.05\%$) for the capsules studied. The change in hydraulic diameters does not affect the Reynolds number and Nusselt number. The entire capsule volume is filled with the storage medium, so there is no air gap inside the capsule. However, it should be noted that, in real applications, the air gap should be left inside the

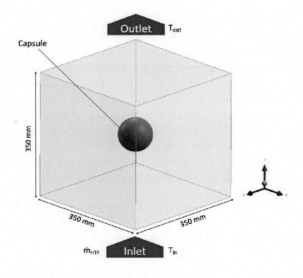

FIG. 5.35

The schematic view of the control volume.

Table 5.18 The View of the Capsules.

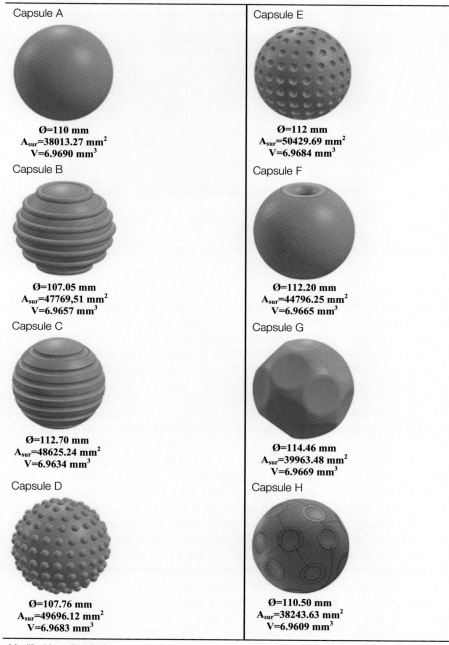

Capsule A	Capsule E
Ø=110 mm A_{sur}=38013.27 mm^2 V=6.9690 mm^3	Ø=112 mm A_{sur}=50429.69 mm^2 V=6.9684 mm^3
Capsule B	Capsule F
Ø=107.05 mm A_{sur}=47769,51 mm^2 V=6.9657 mm^3	Ø=112.20 mm A_{sur}=44796.25 mm^2 V=6.9665 mm^3
Capsule C	Capsule G
Ø=112.70 mm A_{sur}=48625.24 mm^2 V=6.9634 mm^3	Ø=114.46 mm A_{sur}=39963.48 mm^2 V=6.9669 mm^3
Capsule D	Capsule H
Ø=107.76 mm A_{sur}=49696.12 mm^2 V=6.9683 mm^3	Ø=110.50 mm A_{sur}=38243.63 mm^2 V=6.9609 mm^3

Modified from Ref. [20].

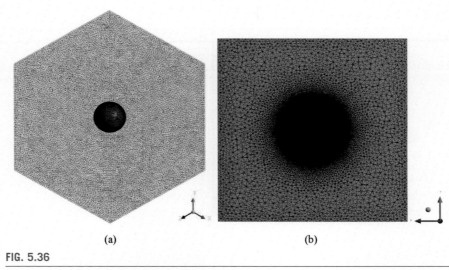

FIG. 5.36

The view of the mesh structure of the control volume: (a) isometric and (b) section view.

Adapted from Ref. [20].

capsule to meet volume changes during phase change. The cases with and without air gap have been compared to determine the effect on the results.

After solid modeling of the capsules and physical domain, it has been meshed by using the mesh module of Ansys Workbench, as shown in Fig. 5.36. The finer mesh has been applied to the zones close to the capsule to obtain a better solution. Grid independence has been applied to the numerical model to reduce the effect of the mesh size on the results. The temperature distribution in the control volume and the variation of the solidification with time have not changed with smaller element size after 2 mm. To provide finer mesh in the zones closer to the contact surface, the contact surface mesh size has been taken 50% lower than the mesh size. For the final numerical model, while the mesh size is 2 mm, and the contact surface mesh size is 1 mm. For capsule A, the number of elements is 4,576,741. The number of elements has varied between 4.5 million and 5.1 million for the other capsules studied.

As the boundary conditions, the bottom surface of the control volume has been set as the velocity inlet, which allows entering velocity, pressure, and temperature data. The upper surface of the control volume has been set as the pressure outlet. Side surfaces of the control volume are defined as adiabatic. The capsule walls have not been modeled as a solid model; they are included in the analyses parametrically as a boundary condition in the contact surface.

Thermophysical features of the working fluids in the analysis are given in Table 5.19, and they are temperature independent. The latent heat of the phase change of water/ice has been taken to be 333.5 kJ/kg, and the phase change temperature is 273.15 K (0°C).

Table 5.19 The Thermophysical Features of the Working Fluids Used in the Analyses.

Substance	c (J/kg K)	ρ (kg/m³)	k (W/m K)	μ (kg/m s)
Water	4182	998.2	0.600	0.001003
Ice	2200	915	0.0454	N/A
Ethylene-glycol 35%	3605	1058	0.465	0.0085

For the numerical calculations, FLUENT 17.1 has been used for solving 3-D continuity, momentum, and energy conversion equations by using the finite volume method. The governing equations have been solved in a pressure-based solver with the COUPLED algorithm employed for the pressure—velocity coupling. The second-order upwind discretization scheme has been used for the convection terms of each governing equation, and PRESTO! has been used as the pressure interpolation scheme. To include the effect of turbulence in the analyses, different turbulence models, which are sea surface temperature (SST) k-omega, transition k-kl-omega, and transition SST, have been tested. Transition k-kl-omega and transition SST models have provided almost the same and the best solution. However, SST k-omega has been applied to the analyses as it provides a faster solution. To model the melting inside the capsule, the solidification and melting module in FLUENT 17.1 code that calculates liquid fraction based on enthalpy balance has been used.

Since analyses have been performed time-dependently, the time step size independence process should be performed to the analysis to eliminate the effect of time step size on the results. Different time step sizes, which are $\Delta t = 0.5$, 0.25, 0.1, 0.05, and 0.01 s, have been applied to the numerical model. The change in the results is negligible when $\Delta t = 0.05$ s and lower. Therefore, the time step size has been set as $\Delta t = 0.05$ s. The gravitational force is included in the analyses with a value of 9.807 m/s^2.

The inlet velocity of the ethylene glycol is tested as 0.001, 0.0025, 0.005, 0.0075, and 0.01 m/s. These velocity values have corresponded to the Reynolds number, which are $Re = 14$, 34, 68, 103, and 137, respectively. The inlet temperature of the ethylene glycol is set as $T_{in} = 2.5$, 5, 7.5, and $10°C$, respectively. At the initial conditions, the temperature of the control volume is $0°C$, and the storage medium is in solid phase (ice).

5.9.2 Validation of the Numerical Model and Procedure

For a numerical analysis, it is required to validate the numerical model and procedure to be sure of the accuracy of the results of the numerical models. For this purpose, the numerical results should be compared with credible experimental results. The results to be obtained from the numerical analyses are not acceptable without validation of a numerical model and procedure. In this case study, the

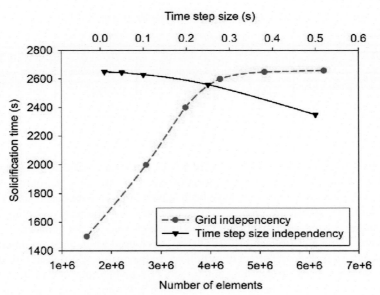

FIG. 5.37

The variations of solidification time with time step size and number of elements.

Data from Ref. [20].

numerical results have been compared with the experimental results. The variation of the solidification time with the number of elements and time step size is shown in Fig. 5.37. The observation regarding grid and time step size independencies discussed previously is clearly shown in Fig. 5.37.

The validation curves for the numerical model are given in Fig. 5.38. The validation of the numerical model and procedure has been performed through the change of volume fraction with time for the storage medium. It is clear from Fig. 5.38 that the numerical model has been achieved to analyze the control volume with precise results. There is a good agreement between experimental and numerical results. In numerical analysis, since the heat gain from the surroundings has been neglected, the melting time is shorter than the experimental study. Root mean square error (RMSE) and mean absolute error (MAE) for the validation process are 0.056 and 0.046, respectively. The effect of the turbulence model of the results is also compared in Fig. 5.38.

5.9.3 Results and Discussion

The melting time or discharging time is an evaluation criterion that provides easy and quick findings to evaluate the performance of the capsules. A longer discharging period means higher thermal performance. The variation of the discharging period with the capsule models is shown in Fig. 5.39. It is clear from Fig. 5.39 that the

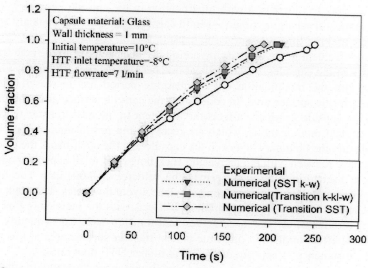

FIG. 5.38

The validation curves for the numerical model.

Data from Ref. [20].

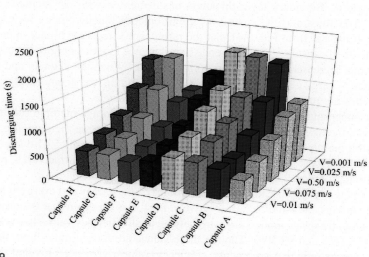

FIG. 5.39

The variation of discharging time for the case of $T_{in} = 283$ K.

Data from Ref. [20].

discharging time has increased with modification in the capsule geometry. Therefore, the shortest discharging time is seen in capsule A, which is a standard sphere. The longest melting time is observed in capsule D.

The improvement in the discharging period has occurred due to the heat transfer surface area of the capsules. Since the heat transfer surface area of the capsules has increased with the modification, the discharging period has increased with the increasing capsule surface area. In addition to the capsule surface area, the modification in the capsule geometry affects the turbulent in the heat transfer fluid flow. The geometrical modifications, which are recesses and protrusions on the capsule surface, disrupt the boundary layers on the capsule. On the other hand, the discharging times have reduced with the increasing HTF flow rates in all capsule models, so the longest discharging period is seen in the velocity of 0.001 m/s. The melting period has decreased with the increasing ethylene glycol flow rates. The main reason for this decrease is the increasing turbulence effects with the increasing flow rates. Increasing turbulence effects cause higher heat transfer rates. Consequently, while the dominant effect on the melting time is heat transfer surface area at lower HTF flow rates, turbulence effects are dominant at higher HTF flow rates.

The velocity distribution for capsules A and D, which are the standard sphere and the best capsule geometry studied, is shown in Fig. 5.40. The difference in the velocity gradient between capsules A and D has increased with the increasing the flow rate of the HTF. The velocity boundary layer is thicker in capsule D than capsule A. Therefore, capsule D provides more extended dead flow area behind it than capsule A. Boundary layer thickness has decreased with the increasing flow rate. Fig. 5.41 illustrates the contour of the turbulence kinetic energy for capsules A and D. It is clear from Fig. 5.41 that the turbulence effects are higher in geometrically modified capsules. The difference in turbulence kinetic energies for capsules A and D has decreased with the increasing flow rate. When the velocity and turbulence kinetic energy contours are taken into consideration, the geometrically modified capsules are more effective in the lower HTF flow rates.

In the encapsulated heat storage systems, when the storage medium solidifies in the capsule, the lower HTF flow rates provide to both decrease the pumping power and prohibit the supercooling effect. It is clear from Figs. 5.40 and 5.41 that geometrically modified heat storage capsules, which have extended surface area, provide longer discharging period than the standard spherical capsule.

The effects of geometrically modified capsules on the energy and exergy values are illustrated in Figs. 5.42 and 5.43, respectively. First, it is observed that the energy efficiencies for all capsule models are extremely high; all efficiency values have changed between 98.15% and 99.50%. Though these values are indeed quite high, they are certainly not expected. Also, energy efficiency values have not provided detailed and comparable information about the capsule models. The energy efficiency is, by definition, the ratio of the desired to the required energy amounts. However, the difference between these two values is the result of viscous dissipation in the fluid. This value, which is analogous to a pressure loss within the fluid, is significant when compared with the thermal energy stored in the water and in the PCM, which explains

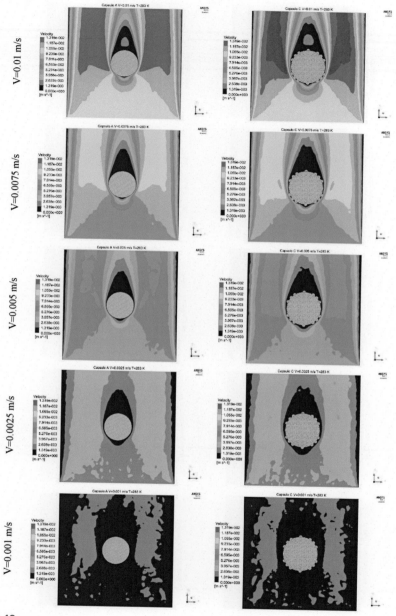

FIG. 5.40

Velocity contours for capsule A and capsule D.

Adapted from Ref. [20].

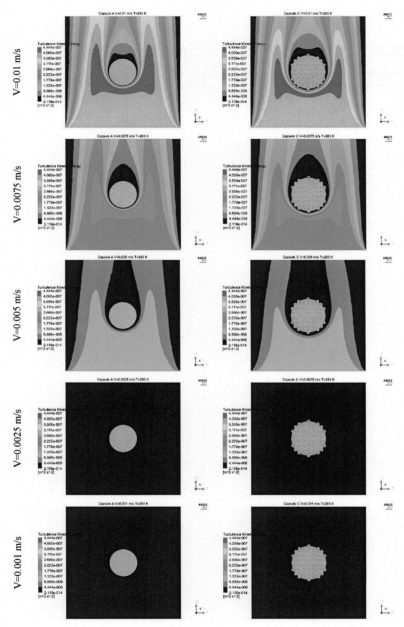

FIG. 5.41

Turbulence kinetic energy contours for capsule A and capsule D.

Adapted from Ref. [20].

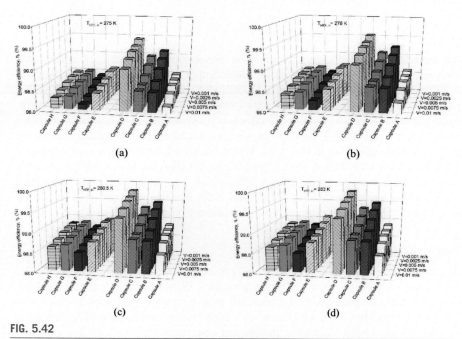

FIG. 5.42

The variation of energy efficiency values with HTF flow rates: (a) $T_{HTF, in} = 275.5$ K, (b) $T_{HTF, in} = 278$ K, (c) $T_{HTF, in} = 280.5$ K, (d) $T_{HTF, in} = 283$ K.

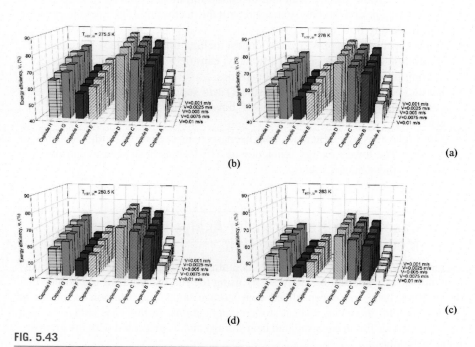

FIG. 5.43

The variation of exergy efficiency values with HTF flow rates: (a) $T_{HTF, in} = 275.5$ K, (b) $T_{HTF, in} = 278$ K, (c) $T_{HTF, in} = 280.5$ K, (d) $T_{HTF, in} = 283$ K.

Data from Ref. [20].

the remarkably high efficiency values. The reason why the normalized energy efficiencies were adopted was that a closer inspection of the differences can be observed.

The exergy analyses are recognized more helpful than energy analysis in a number of ways since they asses both the quantity and the usefulness of energy. In other words, exergy analysis allows examining the effects of irreversibility in the system in addition to the interactions passing through the system boundaries. Exergy efficiencies for all capsule models are given in Fig. 5.42. When energy efficiency values are compared with the exergy efficiencies, the exergy efficiency values are lower and significantly comparable, in contrast to the energy efficiencies. The higher HTF flow rates have increased the exergy destruction. The lower HTF inlet temperatures, which is closer to the phase change temperature, have provided a lower exergy destruction.

5.9.4 Closure

This case study presents a numerical study to determine the effect of geometrically modified heat storage capsules. Seven geometrically modified capsules have been analyzed and compared with the standard sphere capsule. A complete numerical analysis method has been introduced as a heat storage application. The following key results have been reached in the present case study:

- Geometrically modified ice capsules have increased the melting time since they extended capsule surface area and increased turbulence effects.
- The melting period has been extended with the decreasing HTF inlet temperature and increasing flow rate.
- While capsule D has provided the longest melting time, capsule A has caused the shortest discharging period. Additionally, capsule F has supplied the shortest melting time among the modified capsules.
- Parallel to the melting time, Capsule D has supplied the highest energy efficiency, while capsule A and capsule F have supplied the lowest energy efficiency.
- The highest exergy efficiency has been seen in capsule D with the value of 85.12%. The lowest exergy efficiency of 51.91% has been seen in capsule A, which is the standard sphere.

5.10 Case Study: Heat Transfer Analysis for a Porous Medium Heat Storage Unit

As mentioned in Section 4.7.4, porous medium modeling is a critical issue as many heat storage systems can be analyzed by using the fundamentals of the porous medium. Rock beds and encapsulated heat storage tanks are common examples of porous medium modeling as heat storage applications. In this case study, the heat transfer analysis for an encapsulated heat storage tank is presented with the improvement techniques.

5.10.1 System Description

Heat transfer and fluid flow characteristics have a significant effect on the thermodynamic performance of the encapsulated heat storage tank. These parameters also directly affect the capacity, size, and cost of the heat storage unit. As in all heat storage tanks, there is thermal stratification in the heat storage units that consist of the porous medium. Enhanced heat storage units in terms of thermal stratification provide higher thermodynamic and economic performance. Therefore, the determination of the temperature distribution inside the tank, which is used to calculate the degree of the thermal stratification, is a significant issue for the heat storage units consisting of the porous medium. It can be obtained with analytical, numerical, and experimental works. Here, an encapsulated ice storage tank has been analytically modeled by using the fundamentals of porous medium modeling based on Refs. [21,22].

Fig. 5.44 demonstrates the schematic view of the encapsulated heat storage tank. In the storage tank, there are thousands of capsules that are comparatively smaller in

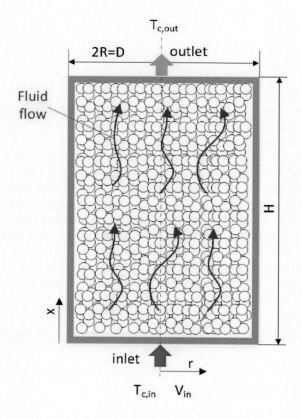

FIG. 5.44

The schematic view of the encapsulated heat storage tank.

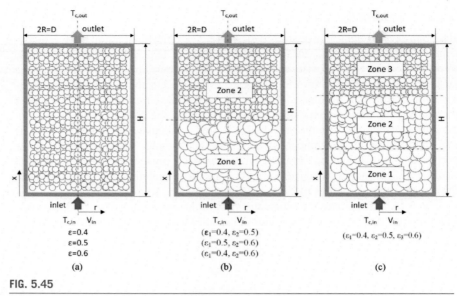

FIG. 5.45

The schematic view of the heat storage tank that consists of graded porosity (a) one-graded structure, (b) two-graded structure, and (c) three-graded structure.

volume than the storage tank. A heat transfer fluid flows through the capsules. The diameter of the tank is D, and its height is H. In the present case study, the diameter and height of the tank are $D = 2$ and 8 m, respectively.

Generally, in the encapsulated heat storage tank, the capsule size is the same throughout the tank. In other words, the porosity of the tank is uniform as shown in Fig. 5.45a. In this case study, to improve the thermal stratification in the heat storage tank, the use of graded porosity use in the tank has been studied. In this context, as shown in Fig. 5.45, the effects of single-, two-, and three-grade porosity in the tank have been investigated. First, the effect of the single porosity is investigated for the cases of $\varepsilon = 0.4, 0.5$, and 0.6. Then, the tank has been divided into two equal volume zones. The cases of $(\varepsilon_1 = 0.4, \varepsilon_2 = 0.5)$, $(\varepsilon_1 = 0.5, \varepsilon_2 = 0.6)$, and $(\varepsilon_1 = 0.4, \varepsilon_2 = 0.6)$ have been modeled. Finally, the tank has been divided into three equal zones: $\varepsilon_1 = 0.4$, $\varepsilon_2 = 0.5$, and $\varepsilon_3 = 0.6$.

5.10.2 Porous Medium Modeling of the Encapsuled Heat Storage Tank

The following assumptions have been applied to the problem to reduce the complexity of the problem:

- The effect of buoyancy forces is neglected.
- The viscous effect is not included in the analysis.
- Heat transfer fluid is incompressible.
- The thermophysical features of working fluids are temperature independent.
- The flow in the axial axis is fully developed.
- The phase change temperature of the storage medium is 0°C. It is also assumed as the temperature of the solid particles in the control volume.

The energy equation in the porous mediums can be written as

$$\varepsilon \rho c u \frac{\partial T}{\partial x} = \frac{1}{r} \frac{\partial}{\partial r}\left(kr\frac{\partial T}{\partial r}\right) + \frac{\partial}{\partial x}\left(k\frac{\partial T}{\partial x}\right) + HA_{bed}\left(T_{pcm} - T\right) \tag{5.33}$$

Eq. (5.33) can be solved easily with the integral method when the temperature changing in the radial axis is assumed as linear or parabolic, which is written as follows:

$$T_r = T_c + \left(\frac{T_0 - T_c}{kR_t + R}\right)r \tag{5.34}$$

$$T_r = T_c + \left(\frac{(T_0 - T_c)}{2kR_T R + R^2}\right)r^2 \tag{5.35}$$

Boundary conditions used to solve Eq. (5.34) are written as

$$T(x, 0) = T_c \tag{5.36}$$

$$\left|\frac{\partial T}{\partial r}\right|_{r=0} = 0 \tag{5.37}$$

$$\left|\frac{\partial T}{\partial r}\right|_{r=R} = \frac{\ddot{q}_w}{k} \tag{5.38}$$

Eq. (5.34) addresses the centerline temperature. Eq. (5.37) is the symmetry condition in the center of the tank. Eq. (5.38) is the boundary condition at the tank wall. It is an energy balance equation that represents the heat conduction inside the tank and heat convection on the outer surface of the tank.

The term of $\ddot{q}_w$ in Eq. (5.38) can be calculated as

$$\ddot{q}_w = \frac{(T_0 - T_w)}{R_T} \tag{5.39}$$

5.10.3 Thermodynamic Analysis

Thermodynamic analysis has been performed based on energy and exergy efficiencies. Energy efficiency of the heat storage tank can be written as

$$\eta = \frac{E_{stored}}{E_{stored} + Q_{gain}} \tag{5.40}$$

Here, E_{stored} can be calculated as

$$E_{stored} = mc\left(T_{c,out} - T_{c,in}\right) \tag{5.41}$$

Q_{gain} denotes the heat gain from surrounding and tank, and it can be calculated as

$$Q_{gain} = \Delta t \int_0^L \left(\frac{T_0 - T_w(x)}{R_T}\right) 2\pi R dx \tag{5.42}$$

Energy efficiency of the heat storage tank can be written as

$$\psi = \frac{Ex_{stored}}{Ex_{in} + Ex_{out}} \tag{5.43}$$

Ex_{stored} can be calculated as

$$Ex_{stored} = Ex_{in} - Ex_{out} + Ex_{Q_{gain}} + Ex_{dest} \tag{5.44}$$

$Ex_{in} - Ex_{out}$ can be written as

$$Ex_{in} - Ex_{out} = mc\left[(T_{c,out} - T_{c,in}) - T_0 ln\left(\frac{T_{c,out}}{T_{c,in}}\right)\right] \tag{5.45}$$

Ex_{loss} can be calculated as

$$Ex_{Q_{gain}} = Q_{gain}\left(1 - \frac{T_0}{T_w}\right) \tag{5.46}$$

5.10.4 Results and Discussion

When Eq. (5.33) is solved by using the boundary conditions given in Eqs. (5.36)–(5.38), with the assumption that the temperature variation in the radial axis is linear, the following equation is reached:

$$T_c(z) = c_1 e^{\lambda_1 x} + c_2 e^{\lambda_2 x} + c_3 \tag{5.47}$$

The terms c_1, c_2, c_3, λ_1, and λ_2 are written as

$$c_1 = \frac{A(B - \varepsilon c_p u_\infty C)}{2BD} \tag{5.48}$$

$$c_2 = \frac{A(B + \varepsilon \rho c_p u_\infty C)}{2BD} \tag{5.49}$$

$$c_3 = \frac{3((T_{PCM} - 2/3T_0)R + kR_T T_{FDM})RA_{bed}H + 6kT_0}{RA_{bed}(3kR_T + R)H + 6k} \tag{5.50}$$

$$\lambda_1 = \frac{CF}{2E} \tag{5.51}$$

$$\lambda_2 = \frac{C - F}{2E} \tag{5.52}$$

The coefficients of A, B, C, D, E, and F can be calculated as

$$A = T_{c,in}HA_{bed}R^2 + 3T_{c,in}HA_{bed}RkR_T + 6T_{c,in}k - 6kT_0 - 3HA_{bed}T_{PCM}RkR_T$$
$$- 3HA_{bed}T_{PCM}R^2 + 2HA_{bed}R^2 T_0 \tag{5.53}$$

$$B = 4R^2HA_{bed}k + \left(Ru_\infty c_p\rho\varepsilon\right)^2 + 3RkR_T\left(u_\infty c_p\rho\varepsilon\right)^2 + 12RHA_{bed}k^2\ R_T + 24k^2 \quad (5.54)$$

$$C = \left[\left(3\varepsilon\rho c_p u_\infty RkR_T\right)^2 + 6RkR_T\left(\varepsilon\rho c_p u_\infty R^2\right)^2 + \left(\varepsilon\rho c_p u_\infty R^2\right)^2 + 72k^3RR_T\right.$$
$$\left. + 4HA_{bed}R^4k + 36HA_{bed}k(RkR_T)^2 + 24k^2R^2 + 24HA_{bed}R^3k^2R_T\right]^{1/2} \quad (5.55)$$

$$D = HA_{bed}R^2 + 3HA_{bed}RkR_T + 6k \quad (5.56)$$

$$E = Rk(3kR_T + R) \quad (5.57)$$

$$F = 3\varepsilon\rho c_p u_\infty RkR_T + \varepsilon\rho c_p u_\infty \quad (5.58)$$

When the temperature variation in the radial axis is assumed as parabolic, Eq. (5.33) is solved as follows:

$$T_c(z) = c_4 e^{\lambda_3 x} + c_5 e^{\lambda_4 x} + c_6 \quad (5.59)$$

The terms c_4, c_5, c_6, λ_3, and λ_4 are written as

$$c_4 = \frac{A\left(B - \varepsilon c_p u_\infty C\right)}{2BD} \quad (5.60)$$

$$c_5 = \frac{A\left(B + \varepsilon\rho c_p u_\infty C\right)}{2BD} \quad (5.61)$$

$$\lambda_3 = \frac{CF}{2E} \quad (5.62)$$

$$\lambda_4 = \frac{C - F}{2E} \quad (5.63)$$

$$c_6 = \frac{2A_{bed}H(T_{PCM} - 0.5T_\infty)R^2 + 4HA_{bed}T_{PCM}RkR_T + 8kT_0}{8k + 4HA_{bed}RkR_T + HAR^2} \quad (5.64)$$

The coefficients of A, B, C, D, E, and F can be calculated as

$$A = T_{c,in}HA_{bed}R^2 + 4T_{c,in}HA_{bed}RkR_T + 8T_g k - 8kT_0 - 4HA_{bed}T_{PCM}RkR_T$$
$$- 2HA_{bed}T_{PCM}R^2 + HA_{bed}R^2T_0 \quad (5.65)$$

$$B = 4R^2HA_{bed}k + \left(Ru_\infty c_p\rho\varepsilon\right)^2 + 4RkR_T\left(u_\infty c_p\rho\varepsilon\right)^2 + 16RHA_{bed}k^2R_T + 32k^2 \quad (5.66)$$

$$C = \left[\left(4\varepsilon\rho c_p u_\infty RkR_T\right)^2 + 8RkR_T\left(\varepsilon\rho c_p u_\infty R^2\right)^2 + \left(\varepsilon\rho c_p u_\infty R^2\right)^2 + 128k^3RR_T\right.$$
$$\left. + 4HA_{bed}R^4k + 64HA_{bed}k(RkR_T)^2 + 32k^2R^2 + 32HA_{bed}R^3k^2R_T\right]^{1/2} \quad (5.67)$$

$$D = HA_{bed}R^2 + 4HA_{bed}RkR_T + 8k \quad (5.68)$$

$$E = Rk(4kR_T + R) \quad (5.69)$$

$$F = 4\varepsilon\rho c_p u_\infty RkR_T + \varepsilon\rho c_p u_\infty \quad (5.70)$$

Eq. (5.33) can be solved as 1−D. 1−D form of Eq. (5.33) can be written as follows:

$$\varepsilon\rho Cu\frac{\partial T}{\partial x} = \frac{1}{r}\frac{\partial}{\partial x}\left(kr\frac{\partial T}{\partial r}\right) + HA_{bed}(T_{PCM} - T) \quad (5.71)$$

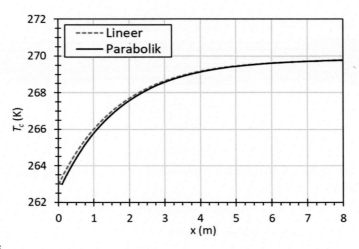

FIG. 5.46

The comparison of the centerline line temperature change for the linear and parabolic assumptions.

Data from Ref. [22].

The effect of the linear and parabolic temperature profile assumptions on the centerline temperature distribution is shown in Fig. 5.46. As seen from Fig. 5.46, there is no significant difference between linear or parabolic assumption. Also, it is possible to obtain the centerline temperature distribution with Eq. (5.71) as a 1-D solution. There is also no significant difference between 1-D and 2-D solutions of the energy equation for the porous medium [21,22].

The centerline temperature variation for all porosity structures is shown in Fig. 5.47 for the case of $u_\infty = 0.001$ m/s and $T_{c,in} = 267$ K. It is clear from Fig. 5.47 that the thermal stratification has improved with reducing porosity for the uniform porosity cases. This occurs because the amount of HTF decreases with the tank with increasing porosity. A lower HTF volume causes a lower energy transported. However, a lower porosity is required to reduce the tank volume. In other words, the heat storage density per volume reduces with lower tank volume at the equal volume of storage medium.

Here, the graded porosity can play a critical role in lowering the tank volume with not reducing the degree of thermal stratification. In the single porosity cases, the centerline temperature distribution is almost the same in the case of $\varepsilon = 0.4$ and $\varepsilon = 0.5$. There is a significant reduction in the case of $\varepsilon = 0.6$. The best centerline temperature distribution is seen in $\varepsilon = 0.4$, which requires the highest tank volume. The enhancement in the thermal stratification with graded porosity is clearly seen in Fig. 5.46. The lower porosity should be located on the bottom side of the tank. The two-zone porosity ($\varepsilon_1 = 0.5$, $\varepsilon_2 = 0.6$) has provided the worst temperature stratification as they are the highest porosities studied. The cases of $\varepsilon = 0.4$, $\varepsilon = 0.5$,

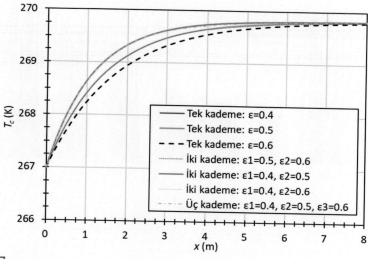

FIG. 5.47

The variation of the centerline temperature with the graded porosity structure for the case of $T_{in} = 267$ K and $u_\infty = 0.001$ m/s.

Data from Ref. [22].

($\varepsilon_1 = 0.4$, $\varepsilon_2 = 0.5$), ($\varepsilon_1 = 0.4$, $\varepsilon_2 = 0.6$), and ($\varepsilon_1 = 0.4$, $\varepsilon_2 = 0.6$, $\varepsilon_3 = 0.6$) have almost same centerline temperature distributions.

To evaluate the porosity profiles in the tank, the thermodynamic analysis based on energy and exergy efficiencies has been performed. Fig. 5.48 demonstrates the energy and exergy efficiencies for all porosity profiles in the case of $T_{c,in} = 267$ K. Energy efficiency values have varied between 95.23% and 99.98%. As seen from these values, energy efficiency values are very close to each other; therefore, they are not providing comparable values. The highest energy values are seen in the cases of $\varepsilon = 0.4$, $\varepsilon = 0.5$, and ($\varepsilon_1 = 0.4$, $\varepsilon_2 = 0.5$, $\varepsilon_3 = 0.6$). As seen from energy efficiency values, three-grade porosity has achieved almost the same energy efficiency performance with the single porosity cases of $\varepsilon = 0.4$ and $\varepsilon = 0.5$. The exergy efficiency values have provided more comparable information, as they have changed between 51% and 85% for all porosity profiles in contrast to the energy efficiency values. The highest exergy efficiency value of 84.71% is seen in the case of $\varepsilon = 0.4$. When the graded porosity profiles are taken into consideration, the highest exergy values are seen in ($\varepsilon_1 = 0.4$, $\varepsilon_2 = 0.5$) and ($\varepsilon_1 = 0.4$, $\varepsilon_2 = 0.5$, $\varepsilon_3 = 0.6$). As seen from exergy efficiency values, the use of graded porosity has achieved the same exergetic performance with the single porosity of $\varepsilon = 0.4$.

Consequently, while the energy efficiency values are very high in all cases, the exergy efficiency values have provided more realistic and comparable results. According to the energy efficiency values, all porosity profiles in the tank have provided almost ideal heat transfer characteristics. In contrast to energy efficiencies, the exergy

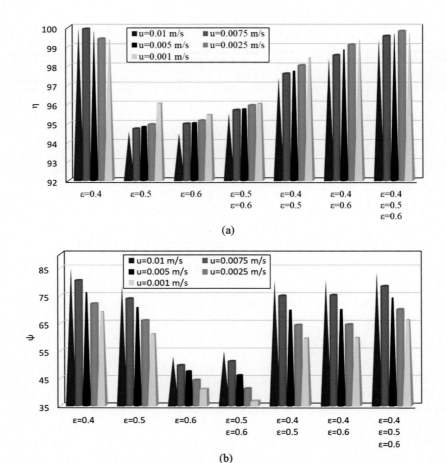

FIG. 5.48

The (a) energy and (b) exergy efficiencies for the case of $T_{c,in} = 267$ K.

efficiency values have provided more realistic and comparable information about the effect of porosity profiles in the tank. The volume of the storage tank directly depends on the porosity of the tank. While a lower porosity tank provides more thermal performance, it requires more volume. To reduce the tank volume, the porosity of the tank should be increased. The graded porosity can help to reduce the volume with minimum thermal performance losses. Thus, it allows storing more energy in the same volume. The size and cost of the storage tank can be reduced significantly.

5.10.5 Closure

This case study presents an application of porous medium modeling in heat storage applications. An encapsulated heat storage tank has been analyzed by solving the energy equation in the porous medium. To minimize the tank volume with minimum thermal performance losses, the application of the graded porosity profiles has been investigated. The following key observations have been reached:

- Lower porosity has provided a higher heat storage performance when a uniform porosity is used in the tank. However, lower porosity requires a higher volume of the tank.
- Graded porosity has provided the same heat storage performance with the uniform low porosity. The lower porosity section should be located on the bottom side of the tank.
- $\varepsilon = 0.4$, ($\varepsilon_1 = 0.4$, $\varepsilon_2 = 0.5$), and ($\varepsilon_1 = 0.4$, $\varepsilon_2 = 0.5$, $\varepsilon_3 = 0.6$) have provided higher thermodynamic efficiencies.
- A lower volume heat storage tank, which has almost the same thermodynamic performance with $\varepsilon = 0.4$, can be achieved with the graded porosity profiles.
- While energy efficiency values have changed between 95.23% and 99.98%, exergy efficiency values have changed between 51% and 85%. The highest exergy efficiency values have been seen in the $\varepsilon = 0.4$, ($\varepsilon_1 = 0.4$, $\varepsilon_2 = 0.5$), and ($\varepsilon_1 = 0.4$, $\varepsilon_2 = 0.5$, $\varepsilon_3 = 0.6$).

5.11 Case Study: Various Heat Storage Application in Buildings

In the previous case studies, various heat storage applications in the buildings have been introduced with the analyzing methods discussed in Chapter 4. In this section, various heat storage methods used in the building application have been introduced briefly. The fundamentals of assessing the systems are almost the same for the systems to be presented. The primary purpose of introducing the applications of heat storage systems is to show the potential of heat storage methods for the readers.

5.11.1 Borehole Heat Storage System

Borehole heat storage systems are used for storing both heating and cooling loads of the buildings, especially for higher capacity storage. The temperature of the soil 10–15 m below the ground does not change significantly for all year. Therefore, this thermal stability of the ground makes it a significant heat source and sinks for the HVAC systems. It is also a significant heat source for ground-based heat pump systems. One of the practical applications of the borehole heat storage is applied in the Ontario Tech University campus. The details of the borehole heat storage system and the method of the thermodynamic analysis have been introduced in Refs. [23–25]. The borehole heat storage system in the Ontario Tech University Campus is used for both heating and cooling demands. There are several buildings in the campus, and during the winter, the cooling demand is quite high due to extreme winter conditions, and during the summer, the heating demand is quite

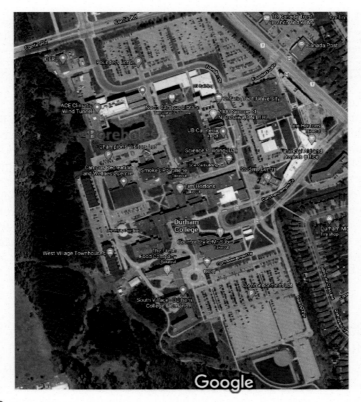

FIG. 5.49

The view of the Ontario Tech Campus site and the zone of boreholes.

high due to hot outdoor temperature and relative humidity. The view of the campus site is shown in Fig. 5.49. As seen from Fig. 5.49, the boreholes are located in the middle of the campus. The cooling and heating demands of the campus buildings are 7000 and 6800 kW, respectively.

The system layout for the borehole heat storage system for cooling purposes is shown in Fig. 5.50. During the summertime, the chillers are operated to reject the heat from the buildings and pump through the boreholes. The ethylene glycol solution is used in the system as the heat transfer fluid. The temperature of the solution at the inlet and outlet sections of the ground piping system is 29.4°C and 35°C, respectively. There is a secondary glycol loop between the system and building fan coils to transfer heat for the cooling of the buildings, which is circulated with the temperature of 5.5°C and 14.4°C, respectively.

The system layout for the borehole heat storage system for heating purposes is shown in Fig. 5.51. During the wintertime, the borehole heat storage system is used for heating the buildings. The inlet and outlet temperatures of the heat transfer fluid are 5.6°C and 9.3°C, respectively. The evaporator water pumps into the

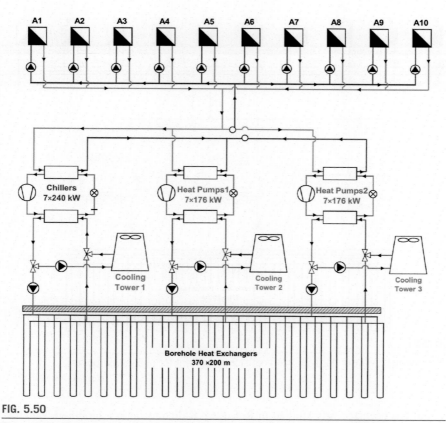

FIG. 5.50

The schematic view of the borehole heat storage system for cooling purposes.

Modified from Ref. [25].

borehole field, and heat is absorbed from the borehole water by the evaporator and transferred to the refrigerant. Heat pumps transfer the thermal energy to the secondary fluid to supply the heat through the buildings. The secondary circulation loop is used for supplying the heat transfer between the heating system and the buildings. The building circulation temperatures are 52°C and 41.3°C, respectively. Also, natural gas furnaces are used for additional heating demands.

Borehole heat storage systems are one of the significant applications of long-term heat storage in buildings. They are easily integrated into renewables. To store renewables for a long term and high capacity, borehole heat storage systems are cost-effective and thermally effective methods.

5.11.2 Heat Storage for Power Generation

Heat storage applications in the buildings are generally used for heating and cooling purposes. On the other hand, there are a few applications of heat storage for power generation. The integrated Kalina cycle with auxiliary heater and heat storage

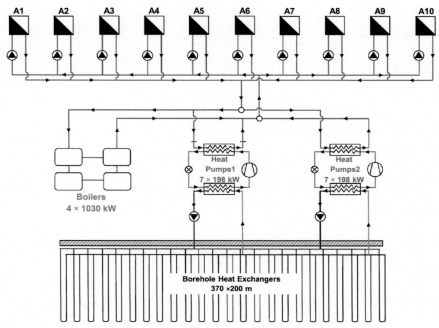

FIG. 5.51

The schematic view of the borehole heat storage system for heating purposes.

Modified from Ref. [25].

system has been recently studied in Ref. [26]. Kalina cycle is a modified Rankine cycle for power generation. A mixture of ammonia water is used as the working fluid instead of pure water. In the present case study, a latent heat storage system has been integrated into the solar-powered Kalina power cycle to continue the power generation during the nighttime. The schematic view of the latent heat storage—integrated solar-powered Kalina cycle is shown in Fig. 5.52.

The system in this case study is in Bandar Aabas, Iran, and the Persian Gulf seawater is used as a coolant for the condenser. A total of 53.85 GWh of heat input is required for a year. 45.13 GWh of the total heat requirement is met by solar energy and stored heat. The remaining heat is met by an auxiliary heater. While the heat requirement is met by solar energy between 07:00 and 19:00, the heat requirement is met by the stored heat and auxiliary heater between 19:01 and 06:59. The results have shown that the efficiency of the Kalina cycle is 6.12%.

The results have shown that the solar-driven Kalina cycle is a significant option to produce power. Heat storage is a significant option to continue the power generation when the sun is down. Heat storage plays an important role to meet the power demands of buildings and continuing power generation at night for the solar-driven Kalina cycle.

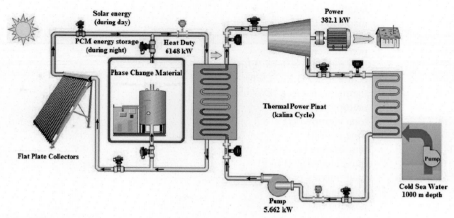

FIG. 5.52

The system layout of the latent heat storage—integrated Kalina cycle for power generation.

Modified from Ref. [26].

5.12 Closing Remarks

In this chapter, the most common heat storage applications used in the buildings are introduced. The advantages and challenges of the heat storage system used in the buildings have been discussed. Results have shown that the heat storage systems should be designed by considering economic, thermodynamic, and heat transfer factors. Heat storage systems play a critical role in managing the grid load, reducing energy costs, and integrating renewables.

Nomenclature

A	area (m^2)
c	specific heat (kJ/kg K)
D	diameter (m)
e	specific energy (kJ/kg)
E	energy (kJ or kWh)
ex	specific exergy (kJ/kg)
Ex	exergy (kJ and kWh)
F	volume fraction of fluid or the view factor for radiation heat transfer
h	enthalpy (kJ/kg) or convective heat transfer coefficient (W/m^2 K)
H	total enthalpy (kJ) or height (m)
k	heat conductivity (W/m K)
K	permeability
KE	kinetic energy (kJ)

L	length (m)
m	mass (kg)
P	pressure (kPa)
PE	potential energy (kJ)
q	heat transfer per area (kJ/m^2)
Q	heat transfer (kJ or kWh)
R	thermal resistance (kW/m^{2}K)
s	entropy (kJ/kg)
S	total entropy (kJ)
t	time (s or h)
T	temperature (K or°C)
u	internal energy (kJ/kg)
U	total internal energy (kJ)
V	volume (m^3)
x, X	position (m)
$\dot{E}$	rate of energy (kW)
$\dot{Ex}$	rate of exergy (kW)
$\dot{m}$	mass flow rate (kg/s)
$\dot{q}$	heat transfer rate per area (kW/m^2)
$\dot{Q}$	rate of heat transfer (kW)
$\dot{S}$	rate of total entropy (kW)
$\dot{V}$	volumetric flow rate (m^3/s)
$\dot{W}$	rate of work (kW)

Greek Letters

α	absorptivity
ε	emissivity
η	energy efficiency
μ	dynamic viscosity (Ns/m^2)
ν	specific volume (m^3/kg) or kinematic viscosity (m^2/s)
ρ	density (kg/m^3) or reflectivity
σ	Stephan–Boltzmann constant
τ	transmissivity
ψ	exergy efficiency
φ	porosity

Subscripts

0	reference conditions
ave	average
b	boundary

boiler	boiler
bottom	bottom of the tank
bulk	bulk temperature of fluid
cap	capsule
ch	charging period
comp	compressor
con	condenser
dest	destruction
disch	discharging period
e	electricity or equivalent
env	environment
eva	evaporator
ex	exergy
f	fluid
final	final
flux	heat flux
gain	gain
gen	generation
HTF	heat transfer fluid
ice	phase of ice
in, inlet	inlet
initial, init	initial
is	isentropic
loss	loss
m	fully mixed
mantle	mantle
net	net
out	outlet
ove	overall
p	constant pressure
pcm	phase change material
pump	pump
R	refrigeration
s	source or solid
sb	solid block
sen	sensible
sf	solid/fluid phase changing
sh	shaft
sm	storage medium
st	storing period
sur	surface
sys	system
$t = 0$	initial conditions of any period
tank	storage tank
top	top of the tank
tur, turbine	turbine
valve	valve

References

[1] D. Erdemir, Determination of effect of bottle arrangement in the sensible thermal energy storage system consisting of water-filled PET bottles on thermal performance, Hittite J. Sci. Eng. 6 (2019) 235–242, https://doi.org/10.17350/HJSE19030000153.

[2] D. Erdemir, N. Altuntop, Thermodynamic analysis of sensible thermal energy storage in water filled PET bottles, Int. J. Exergy 26 (2018), https://doi.org/10.1504/IJEX.2018.092507.

[3] N. Altuntop, Y. Tekin, D. Demiral, Analytical investigation of the use of water filled P.E.T. bottles as thermal energy storage unit, in: 4th Int. Therm. Energy Congr., 2001.

[4] N. Altuntop, Y. Tekin, The first performance results of the solar heating and thermal energy storage by using P.E.T. bottles, in: Eur. Sol. Congr. (EUROSUN 2002), 2002.

[5] D. Erdemir, N. Altuntop, Improved thermal stratification with obstacles placed inside the vertical mantled hot water tanks, Appl. Therm. Eng. 100 (2016), https://doi.org/10.1016/j.applthermaleng.2016.01.069.

[6] D. Erdemir, N. Altuntop, Effect of thermal stratification on energy and exergy in vertical mantled heat exchanger, Int. J. Exergy 20 (2016), https://doi.org/10.1504/IJEX.2016.076681.

[7] D. Erdemir, H. Atesoglu, N. Altuntop, Experimental investigation on enhancement of thermal performance with obstacle placing in the horizontal hot water tank used in solar domestic hot water system, Renew. Energy 138 (2019), https://doi.org/10.1016/j.renene.2019.01.075.

[8] D. Erdemir, N. Altuntop, Experimental investigation of phase change material utilisation inside the horizontal mantled hot water tank, Int. J. Exergy 31 (2020), https://doi.org/10.1504/IJEX.2020.104722.

[9] D. Erdemir, Experimental and Numerical Investigation of the Effects of Using Phase-Changing Materials to Increase Heat Storage Capacity in Vertical Mantled Hot Water Tanks and Their Performance in Different Operating Conditions, Final Project Report, 217M993, 2019.

[10] D. Erdemir, A. Ozbekler, N. Altuntop, Experimental investigation on the effect of the ratio of tank volume to total capsulized paraffin volume on hot water output for a mantled hot water tank, Sol. Energy (2021).

[11] Google, Gimsa Supermarket, Eryaman, 2021 (n.d.), https://www.google.com/maps/@39.9752672,32.6044471,17.96z. (Accessed 20 September 2021).

[12] D. Erdemir, N. Altuntop, Y.A. Cengel, Experimental investigation on the effect of ice storage system on electricity consumption cost for a hypermarket, Energy Build. (2021).

[13] Turkish State Meteorological Service, 2019. https://mgm.gov.tr/eng/forecast-cities.aspx?m=ANKARA. (Accessed 2 October 2019).

[14] Cryogel, Ice Ball Thermal Energy Storage, 2019. http://www.cryogel.com/. (Accessed 27 December 2019).

[15] D. Erdemir, N. Altuntop, Effect of encapsulated ice thermal storage system on cooling cost for a hypermarket, Int. J. Energy Res. (2018), https://doi.org/10.1002/er.3971.

[16] N. Eskin, H. Türkmen, Analysis of annual heating and cooling energy requirements for office buildings in different climates in Turkey, Energy Build. 40 (2008) 763–773, https://doi.org/10.1016/j.enbuild.2007.05.008.

[17] M. Bulut, Thermodynamic and Economic Investigation of the Use of Ice Thermal Energy Storage Systems as the Heat Source in a Heat Pump, Erciyes University, 2021.

[18] D. MacPhee, I. Dincer, A. Beyene, Numerical simulation and exergetic performance assessment of charging process in encapsulated ice thermal energy storage system, Energy 41 (2012) 491–498, https://doi.org/10.1016/J.ENERGY.2012.02.042.

[19] R.I. ElGhnam, R.A. Abdelaziz, M.H. Sakr, H.E. Abdelrhman, An experimental study of freezing and melting of water inside spherical capsules used in thermal energy storage systems, Ain Shams Eng. J. 3 (2012) 33–48, https://doi.org/10.1016/j.asej.2011.10.004.

[20] D. Erdemir, Numerical investigation of thermal performance of geometrically modified spherical ice capsules during the discharging period, Int. J. Energy Res. 43 (2019), https://doi.org/10.1002/er.4585.

[21] D. MacPhee, I. Dincer, Thermal modeling of a packed bed thermal energy storage system during charging, Appl. Therm. Eng. 29 (2009) 695–705, https://doi.org/10.1016/j.applthermaleng.2008.03.041.

[22] D. Erdemir, N. Altuntop, Investigation of the effect of using graded porosity on the thermal performance of the encapsulated ice thermal energy storage tank, Energy Storage 1 (2019) e44, https://doi.org/10.1002/est2.44.

[23] I. Dinçer, M.A. Rosen, Thermal Energy Storage: Systems and Applications, second ed., 2010, https://doi.org/10.1002/9780470970751.

[24] O. Kizilkan, I. Dincer, Borehole thermal energy storage system for heating applications: thermodynamic performance assessment, Energy Convers. Manag. 90 (2015) 53–61, https://doi.org/10.1016/j.enconman.2014.10.043.

[25] O. Kizilkan, I. Dincer, Exergy analysis of borehole thermal energy storage system for building cooling applications, Energy Build. 49 (2012) 568–574, https://doi.org/10.1016/j.enbuild.2012.03.013.

[26] M. Mehrpooya, B. Ghorbani, S.A. Mousavi, Integrated power generation cycle (Kalina cycle) with auxiliary heater and PCM energy storage, Energy Convers. Manag. 177 (2018) 453–467, https://doi.org/10.1016/j.enconman.2018.10.002.

Artificial Intelligence in Heat Storage Applications

6.1 Introduction

Today, artificial intelligence (AI) is recognized as a potential area of science, engineering and technology and expected to play an essential part of our lives. The future of people will depend much on AI technologies and their immediate implementations. We directly use AI-managed devices and provide input data for AI systems. For example, translation robots and applications are commonly used by people, and they solve communication problems instantly. While using it to meet our demands, we provide the input data for training and validation for improving themselves. Virtual assistants are another example of AI application in our daily life. They listen to us, answer our questions, and recommend some activities according to our habits. Every day, they increase the number of works to able to do and do them better. Robot vacuums and mops are other examples of AI technologies from daily life. They can map places and clean them up in the shortest way without skipping any space. Consequently, we are data providers of AI systems as much as we are users. In addition to these common examples of AI, there are many AI applications for numerous purposes in various sectors. For example, AI software is used in the diagnosis processes, treatment, and surgery planning procedures. They can optimize the production planning procedures for a factory. The applications of AI, of course, are not limited to these examples.

Although there are many methods to develop AI models, one of the most critical issues is to find the data for the training, test, and validation processes of AI. Data for AI can be obtained theoretically or can be obtained from experimental or numerical data. When the theoretical model is used, the analytic or empirical equation sets are entered into the AI model. Thus, the AI can generate the result data for different cases regarding the model virtualized. Also, AI can update the independent variables and constants in the equation sets to improve the accuracy of the model. This aspect of AI can be used for the optimization of physical systems. On the other hand, by using the input and output data obtained from experimental and numerical works, an AI model can be developed. Thus, it is possible to investigate how different scenarios, cases, or input data affect the result. This method plays a critical role in cases that are not possible to build a theoretical model for the problem studied.

As in every sector, AI applications have been receiving increasing attention by many, including researchers, scientists, engineers, and technologists from academia to industry. They can be applied to numerous practical energy conversion, storage, and generation systems. As mention previously in Section 1.3, the application of AI into energy systems is one of the branches of the smart energy portfolio as introduced by Dincer [1]. To benefit from an energy source in the most effective way, the subsystems used in the main integrated energy system should be managed and controlled intelligently. Today, when the complexity of the integrated energy systems is considered, the integration of AI into them is an essential need. With the AI models, the energy systems can be designed in a shorter time, their economic performance can be increased, the subsystems and also entire systems can operate more efficiently, and lastly, manpower to be used for designing and operating the system can be reduced significantly.

AI applications can be integrated into heat storage systems in many ways. For example, AI can be used for determining the performance of heat storage systems in different climatic conditions, as the climatic conditions directly affect the performance of the heat storage systems. Many AI models have been introduced for determining the heating and cooling load of buildings in different climatic conditions. Also, AI can be used for predicting future cases in the long and short terms. There are numerous AI applications regarding how meteorological data will change in the future. On the other hand, it is also used for managing the subsystems according to the demand or supply profiles in the short term. For example, the use of the stored heat can be extended, or the charging period can be shorted according to current outdoor conditions.

As mentioned earlier, numerous AI models and their applications into physical problems are presented in the literature. Here, we are focusing on the heat storage— related applications of AI. A brief background regarding AI is first introduced. Then, the basics of the artificial neural network (ANN) are given. Finally, applications of AI in heat storage systems are discussed with the case studies.

6.2 Artificial Intelligence

AI is a field of study that aims to understand intelligence and create intelligent beings. The aim of AI technologies is often defined as hominoid thinking/behavior and sensible thinking/behavior.

Hominoid behavior is generally defined through the Turing test, which is a technique of inquiry in AI to decide to have or not have capable of thinking like a human being. The test is named after Alan Turing, an English computer scientist. In the test, a human questioner asks questions to a computer and a human. The questioner tries to decide which respondent is human and which is a computer. If the questioner makes the correct estimation in half of the tests or less, the computer is considered to have AI. To pass the Turing test, an AI computer should have the following features:

- Natural language processing: A computer should able to communicate with the languages people talk.
- Knowledge representation: It should able to store the knowledge it owns or gathers.
- Automated reasoning: It should able to respond to the questions with the knowledge it has and to produce new responses by using the knowledge it has.
- Machine learning: It should able to adapt itself to the new rules and conditions and to estimate new problems.

The Turing test does not consider people's physical interaction such as vision, hearing, touch, and so on. However, today, people's physical interactions are essential part of the AI technologies. Therefore, the following should be included in the AI criteria:

- Computer vision
- Robotics

Humanoid thinking of AIs stands for computer's ability to think like a human. Sensible thinking of AIs is that it aims to reach the correct answer based on the answer it has. Humanoid behavior of AIs is to control the answer generated by computers whether it is suitable for human behavior. AI technologies are integrated into many applications from almost every sector. They are essential part of the daily life, as many devices managed by AI are used in daily life.

AI can be used in every stage of heat storage applications, from design to operation of them. For example, climatic conditions are crucial for heat storage systems, and they directly affect performance, designs, and costs. There are many AI-aided software packages to estimate the climatic conditions in a location. The software generates data for the location considered with high accuracy. Those data are used as input data for the heat storage system. Additionally, many AI-based software packages are used to estimate the future climatic conditions based on historical data. As another example, the AI can be used to predict values significant for the heat storage system, such as temperature gradient, supply and demand profiles, and so on. Also, they can include the systems for operating the heat storage system more effectively. Lastly, it is a crucial part of the optimization of heat storage systems. They can be used for maximizing the benefits to be obtained from heat storage systems or minimizing the initial costs, equipment sizes, etc.

6.3 Artificial Neural Network

ANN is a subsection formed under the concept of AI and has become the focus of researchers interested in this subject. ANNs collect information from examples, generalize inferences by using those examples, and then make decisions about problems that they have never seen using the information they learn from the examples. Due to these learning and generalization features, ANNs find wide application opportunities in many sectoral fields today and have the ability to successfully solve complex problems with very low errors.

ANN studies have first started with the modeling of neurons, which are biological units of the brain. ANN can be defined as a system designed to model the way the brain performs a function. ANNs are computer programs that mimic biological neural networks. They have a parallel and distributed information processing structure, inspired by the human brain. They consist of processing elements that are connected to each other through weighted connections and each having its own memory. Fig. 6.1 demonstrates the comparison of the human neural system and ANN. The working principles of the human neural system and ANN are almost the same. The neuron in the human is the processor element in the ANN. Dendrite is the collection function. The cell body is the transfer function. Axons are the output of the ANN. The synapses are the weights in ANN.

The main difference between the human neural system and ANN is that the human neural system may forget, but ANNs cannot. A fully trained, tested, and

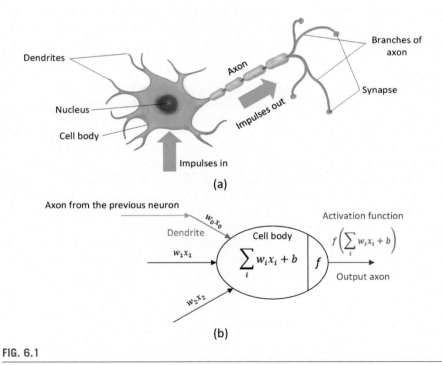

FIG. 6.1

The comparison of the (a) human neural system and (b) ANN. *ANN*, artificial neural network.

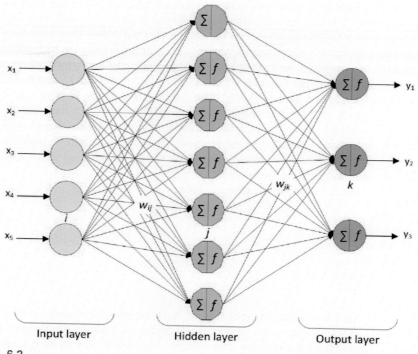

FIG. 6.2

The diagram of the artificial neural network.

validated ANN does not forget. Also, the processing speed is another difference. After full training, an ANN can be faster than a human neural network system. Accuracy is another difference between the human neural system and ANN. While an ANN always generates the same information once it is fully trained, the human neural system does not. It can generate different results at the same input conditions.

An ANN consists of connecting artificial nerve cells with each other in various ways, which are arranged in layers, as shown in Fig. 6.2. Artificial nerves need not be connected from all inputs. Elements in the input and output layers are not neurons, as there are no transfer and activation functions at the same time. The lines between the elements and cells denote the weights (w_{ij} and w_{jk}), which vary between 0 and 1. While 1 represents the most powerful connection, 0 denotes the weak connection. The basic ANN consists of three layers: input, hidden, and output layers.

The input layer is the layer where inputs from the external world come to the ANN. In the input layer, the input data are directly sent to the hidden layer. The number of neurons in the input layer is equal to the number of input parameters in the problem. In the hidden layer, the data received from the neurons in the input layer or the previous hidden layer get transmitted to the next layer by processing with activation functions. The number of hidden layers and the number of neurons in the hidden layers have a significant impact on the accuracy of the ANN. They can change from network to network according to the complexity of the problem to be modeled. The number of neurons in the hidden layers is independent of the number of inputs and outputs. Therefore, the number of hidden layers and the number of neurons in them should be studied parametrically. The output layer is the response of the ANN to the problem. The data coming from the hidden layer get transmitted to the external world by processing with the activation function. The number of output neurons is determined according to the problem. However, it should be noted that every ANN should have one output. Each neuron in the output layer is connected to all neurons in the previous layer.

The ANN has the following features to offer:

- learning,
- attribution,
- classification,
- generalization,
- characterization, and
- optimization.

Note that ANN is not used in problems that have the exact mathematical model. However, it can be used for a special purpose. It creates its personal responses with the facts collected from the samples after making comparable choices on comparable issues. It can be used very effectively in cases where there is no information about events, but examples are available. On the other hand, ANNs have some disadvantages. They work depending on hardware. The pattern of the ANN has a great impact on the precise results. Showing the problem that the network will learn is an important problem. ANN only works with numerical values. Therefore, the problem must be transformed into numerical values. There is no certain criterion for how much the network needs to be trained. Reducing the error of the network on the samples below a certain value is considered sufficient for the completion of the training. Despite these disadvantages, the ANN is used for modeling many problems from various sectors. The ANN can be used for the problems that have the following criteria:

- nonlinear,
- multidimensional,
- noisy,
- complex,
- uncertain,
- missing data,
- defective,
- without mathematical model, and
- without algorithm.

The ANN may generally be used for various purposes, including but not limited to the following:

- probabilistic function estimates,
- intelligent and nonlinear control,
- nonlinear signal processing,
- nonlinear system modeling,
- pattern recognition,
- optimization,
- attribution or shape matching,
- signal filtering, and
- classification.

Note that the reference sources should be consulted for detailed information related to the fundamentals of the ANN. There are many modeling algorithms, activation functions, and patterns. They should be selected accordingly to the specific problems. Today, many powerful software and submodules under the main software are available to create patterns and train, test, and validated the ANNs.

6.4 Artificial Neural Network Applications in Energy Systems

ANN is expected to help to design, operate, and evaluate the energy systems effectively. Energy systems are generally designed specifically for the location and conditions they build. Many parameters affect their design such as the climatic condition of the surrounding, energy source, energy production capacity, physical condition of the facility, and so on. Generally, while the design processes of the energy systems are performed, each parameter, or a few of them simultaneously, is taken into consideration separately. However, all of them are not taken into consideration simultaneously. The main reason for this is that analyzing many parameters at the same time is more complicated and harder to solve. At this point, the ANN can play a significant role to determine the relation between the parameters and optimize the system better. Also, design criteria can be adjusted according to different climatic conditions or system capacities easily with ANNs. In the open literature, there are many data regarding energy system that are ready to use for ANN modeling.

From the past to present, the ANNs are used for many applications in energy systems. In the following subsections, the applications of ANNs in various energy systems are introduced to the readers.

6.4.1 Artificial Neural Network Applications for Meteorological Data

Meteorological conditions have significant impact on almost all energy systems. Therefore, they should be carefully compiled for the location where the energy systems to be set up, especially for solar and wind systems. On the other hand, the meteorological data are not always available for every location. The expensive experimental setups and long-term experimental procedures are required for gathering meteorological statistics.

Weather forecasting has become a significant research topic for many sectors. The most forecasting methods were using a linear relation between input and output data. However, the relation between input and corresponding data is nonlinear for weather forecasting. Therefore, the ANN can play a critical role due to this no linearity. Abhishek et al. [2] have developed the ANN model to forecast the weather data with effective and accurate nonlinear predictive models. They have also compared the performance of the ANN model developed that uses a different number of hidden layers and neurons, and different transfer functions. In their ANN model, there are 10 input cells in the input layer. They have tested up to five hidden layers that have 16 neurons. They have achieved developing an accurate ANN for estimating the maximum temperature.

Water volumes such as lake, river, sea, and so on are commonly used in the energy systems for various purposes such as cooling, heat source for heat pumps, or evaporating for cryogenic gases. Therefore, the temperature of the water volume is critical for the energy systems. On the other hand, the water temperature depends on many parameters such as weather temperatures, wind speed, humidity, and so on. There is a nonlinear relation between these parameters and water temperature. Temizyurek and Celik [3] have aimed to develop an ANN model to determine the effects of meteorological parameters on water temperatures at Kızılırmak River in Turkey. Air temperature, wind speed, relative humidity, and previous water temperatures have been as input parameters for estimating the water temperature. The data set is from 1995 to 2007. The activation functions, the number of neurons, and hidden layers have been determined by trial-and-error method to find the best results. Fig. 6.3 demonstrates the comparison of experimental data and predicted data with ANN. They have concluded that the meteorological data can be used to simulate water temperature with ANN model for Kızılırmak River.

Rajendra et al. [4] have developed an ANN model to predict meteorological data by using multiple-layer perception and radial base function. Air temperature, relative humidity, and soil temperature have been used as input parameters, which are obtained from India Meteorological Department. There are two hidden layers with four neurons. The results have validated with data obtained by multilinear regression model, recorded by meteorological stations. The ANN models have achieved between 91% and 96% accuracy for predictions of all cases. This work has presented that the use of ANN to predict the meteorological data is a reliable way.

It is clear that meteorological data are the significant parameter for energy systems: both system design parameters and performances. However, it is not possible to get meteorological data for every location. ANN has the capability to estimate the meteorological data for these locations. Also, it is possible to predict the future weather forecast. Lastly, the parameters that are affected by meteorological data, such as water and soil temperatures, can be determined with a lower error.

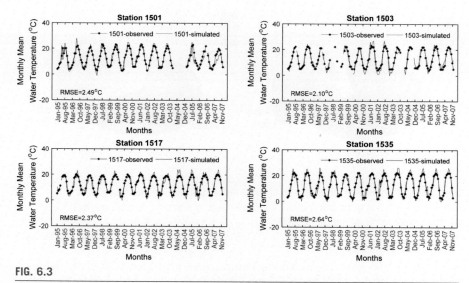

FIG. 6.3

The comparison of experimental data and predicted data with the ANN. *ANN*, artificial neural network.

Adapted from Ref. [4].

6.4.2 Artificial Neural Network Applications for Renewables

Renewables are the key point for carbon-free future and hydrogen economy. They have been included to energy systems for better sustainability. Renewable energy sources directly depend on meteorological and geological data. The relation between these parameters and energy generation rates is nonlinear and hard-to-model mathematically. Although there are many correlations and empirical equations to estimate their performance and design parameters, they are not effective to estimate them.

Ghritlahre et al. [5] have reviewed more than 70 research papers regarding ANN applications of solar air heaters. Fig. 6.4 shows that while the MLP neural model is used for prediction, Levenberg—Marquardt is generally used as a training algorithm. It is clear from these results that ANN method is an appropriate method for performance prediction of different types of solar air heaters. ANN provides more accurate information with reduced computational time.

A few prediction methods to wind speed, which is the most important parameter of the wind energy, have been published on the literature. Lawan et al. [6] have developed an ANN model for prediction of wind speed in the areas where wind speeds velocity does not exist. As the input parameters, a total of 10 years data from five locations between 2003 and 2012, topographical parameters (latitude, longitude, and elevation), and meteorological data have been used for the network training, testing, and validation. Totally, there are seven input cells and one output neuron, which is wind speed in the model. Feed-forward backpropagation is the

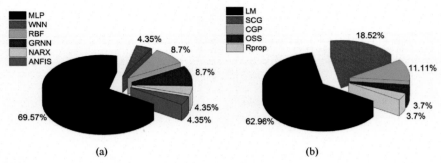

FIG. 6.4

Summary of the present work related to ANN technique for solar air heaters. (a) Percentage of types of neural model. (b) Percentage of learning algorithms used for training. *ANN*, artificial neural network.

Adapted from Ref. [5].

ANN model, and gradient descent backpropagation algorithm is used for training. The accuracy of the ANN has been done through mean square error. Log-sigmoid function is the transfer function in the model. Fig. 6.5 shows the training performance of the ANN. The ANN model has achieved 96% accuracy.

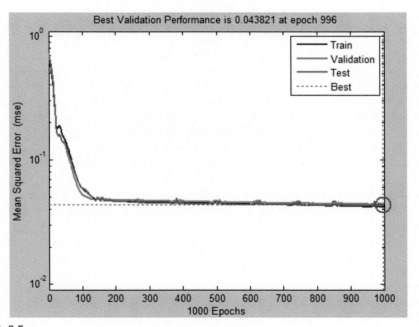

FIG. 6.5

The training performance of the ANN. *ANN*, artificial neural network.

Adapted from Ref. [6].

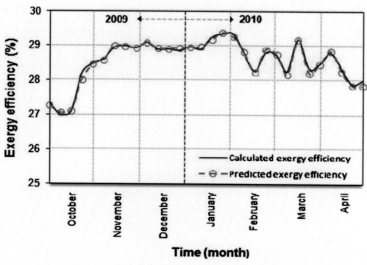

FIG. 6.6

The comparison of the predicted exergy values by the ANN and calculated exergy values. *ANN*, artificial neural network.

Adapted from Ref. [7].

The ANN modeling for estimating the exergy efficiency of a geothermal district heating system has been introduced by Kecebas et al. [7]. This case study is from Afyonkarahisar geothermal district heating system in Turkey. The average daily actual thermal data acquired from the Afyonkarahisar geothermal district heating system in the 2009—10 heating season are acquired and applied for exergy analysis. The ANN has been modeled based on a backpropagation learning algorithm for estimating the exergy efficiency.

Note that the ANN model has three layers. A feed-forward model with tangent sigmoid activation function is selected as the ANN model. The Levenberg—Marquardt backpropagation algorithm has been used in the training. There are five input parameters corresponding to one output. Fig. 6.6 illustrates the comparison of the predicted exergy values by the ANN and calculated exergy values. The results present that the developed ANN has achieved to predict the exergy efficiency values with a maximum correlation coefficient and minimum root mean square values.

6.4.3 Artificial Neural Network Applications for Power Management

The mismatch between power supply and demand profiles is one of the biggest problems in the energy sector. The energy systems are generally designed for the peak demand. The main issue is that the demand-based design is the fluctuating load distribution of energy consumption. Fig. 6.7 shows the electricity demand profile of the

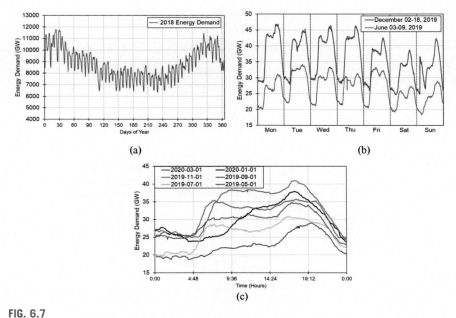

FIG. 6.7

The electricity demand profile of the United Kingdom: (a) daily, (b) weekly, and (c) yearly bases.

Data from Ref. [8].

United Kingdom (Great Britain) as daily, weekly, and yearly bases. It is clear from Fig. 6.7c that while the difference between peak and minimum demands is 80%, it is 30% for the peak and average demands in a daily base. The similar trend is seen in monthly and yearly bases, too. Also, it should be emphasized that the peak demands are seen for a limited period of time, but varying drastically. Since the high-capacity devices are required for meeting the peak load, it is costly to meet. In other words, additional energy devices/plants or energy inputs are required for meeting peak loads, which are seen in limited period of time, so it is costlier.

In addition to the fluctuating behavior of the energy demand, it has been growing every year due to new devices coming into our lives. Here, there are two main problems:

1. The effective demand-based design of the energy systems
2. Predicting how to change energy demand profile with the new items increasing it

These two problems are hard-to-model mathematically, and there is a nonlinear relation between the dependent and independent parameters. Here, the ANN is a convenient tool to model the energy demand profiles. In the open literature, there are many works investigating the energy demand profiles and energy demand-based energy systems. Today, despite there is a limited practical application

Table 6.1 The Validation of the ANN Model Developed by Macedo et al. [9].

Type	Mean Error	Maximum Error	Success (%)
0	0.0124	0.0353	96.4
1	0.0260	0.0362	94.5
2	0.0179	0.0269	98.2
3	0.0104	0.0153	100

ANN, *artificial neural network.*

of them, in the future, these systems will be an inevitable part of the energy systems used in many systems from various sectors. Thanks to these AI-powered devices, the energy source will be used more effectively, with a lower cost and lowering carbon emissions.

Smart grid management is today a global need for energy supply and demand profiles. It provides effective grid management with the use of data generated by dynamic networks. The biggest issue in the smart grid management is to transform the numerous volumes of data into useful data and information for the energy sector. As mentioned previously, almost all energy systems are designed according to demand profile. Therefore, demand-side management methods are an example of smart grid management to optimize the management of the power systems. Macedo et al. [9] have developed an ANN model using data obtained by the use of demand-side management. In the present case study, they have gathered data acquired from digital meters to obtain load curves. The ANN has been trained, tested, and validated with these data. The ANN developed has been used for classifying new data coming from the digital meters and load curves. The data sets have been obtained from 2000 random low-voltage users from various sectors and by a local energy distribution company. The weekend data have been evaluated separately as they are different from the weekday data. The MATLAB ANN Toolbox has been used for modeling ANN. The ANN has three layers: input, hidden, and output layers. Input, hidden, and output layers have 10, 20, and 1 neurons, respectively. Hyperbolic tangent function has been used as the activation function. During the training, the backpropagation algorithm has provided the best solution. To validate the ANN, a simulation has been done with different 220 data sets. Table 6.1 shows the error calculated from the difference between the actual value and the predicted value. It is clear from Table 6.1 that the ANN developed has classified the sample accurately.

6.4.4 Artificial Neural Network Applications for Building Energy Systems

Many analytical, empirical, and numerical methods are used in calculation and prediction of the energy supply and demand profiles in the buildings. While analytical and empirical methods work with rough sensitivity, the numerical methods require high calculation time and powerful computers. Here, ANN can play a significant role in modeling the energy supply and demand profiles in the buildings since there are many data sets available in the open literature. There are two main challenges in the ANN application in the energy systems of buildings:

1. To generate proper data sets for ANN from the existing data
2. To develop accurate ANN models for practical problems

Yalcintas and Akkurt [10] have developed an ANN model to estimate chiller energy use in buildings wherein there are tropical climates with minor seasonal and daily variations. A 42-story commercial building with 41,800 m^2 space in downtown Honolulu, Hawaii, has been used in the case study. The power consumptions of the AC plants have been modeled with the ANN. As input climatic data, dry bulb temperature, wet bulb temperature, dew point temperature, relative humidity percentage, wind speed, and wind direction have been used. As the system works in a time frame, time is also an input parameter. The output of the ANN is the power consumption of AC. The ANN model developed has achieved the predicted energy consumption of the AC used in a tropical climate.

Green roof, also called as eco roof and living roof, is a type of roof that covered a layer of soil on the roof area to grow plant. Green roofs have main two impacts on the buildings.

1. They act as an insulation layer.
2. They remove the heat from the buildings and reduce the cooling loads.

Although the green roofs are used naturally in ancient times, today, they are a commercial product that helps to enhance thermal comfort and air-conditioning. Therefore, it is critical to determine the performance of the green roofs. The performance of the green roofs is determined by experimental or theoretical works. Experimental studies in real condition require long time periods and working in different locations. Laboratory-scale experimental studies have limitation on simulating the environment. Also, the theoretical energy and heat transfer models of green roof are complicated. Here, ANN can play a critical role to estimate the performance of the green roofs.

Erdemir and Ayata [11] have developed an ANN model to predict the temperature decrease provided by green roof in different climatic conditions. While time, total solar radiation, air temperature, wind speed, and relative humidity are used as input parameters, the temperature decrease is used as the output parameter. Fig. 6.8 shows the input parameters used in the ANN. The total 216 data sets have been used for ANN modeling. The data sets have been distributed randomly for the training, testing, and validation with the ratio of 70%, 15%, and 15%, respectively. The schematic of the ANN model is shown in Fig. 6.9. The ANN consists of three layers: input, hidden, and output layers.

The ANN model validation has been performed over the comparison of the actual temperature decrease and predicted ones for nine cities used in the ANN modeling processes. Fig. 6.10 demonstrates the comparison of the temperature decreases in the nine cities. As seen from Fig. 6.10, there is a good agreement between actual and predicted values.

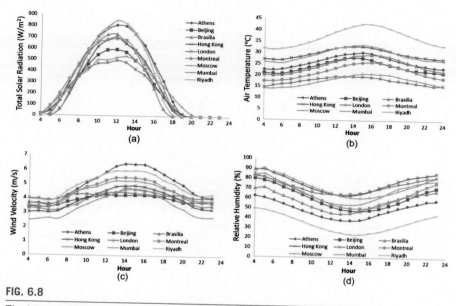

FIG. 6.8

The input parameter used in the ANN modeling (a) total solar radiation, (b) air temperature, (c) wind velocity and (d) relative humidity. *ANN*, artificial neural network.

Data from Ref. [11].

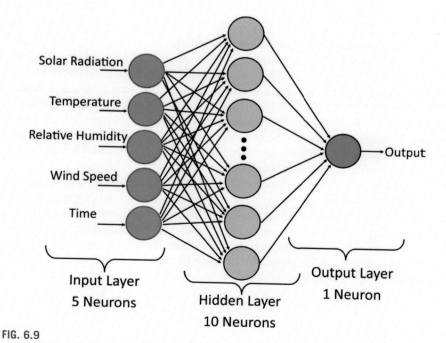

FIG. 6.9

The ANN model used in the green roof modeling. *ANN*, artificial neural network.

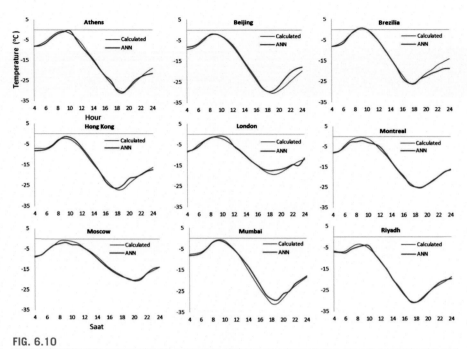

FIG. 6.10

The validation of the ANN model. *ANN*, artificial neural network.

Adapted from Ref. [11].

The ANN model developed has been used for estimating the temperature decreasing values for different cities in Turkey. ANN model has given sufficient result with 99.7777% R and 0.3982% RMSE value. Temperature decreases on the roof for nine cities from Turkey are shown in Fig. 6.11. As seen from ANN results for the cities from Turkey, the green roof has achieved the reduced roof temperature significantly. While temperature decreasing is lower on the roof in the morning, temperature decreasing is higher in the noon and afternoon, as it is expected. Minimum temperature decreasing has been measured between 06:00 and 10:00. The maximum temperature decreasing has been occurred at 18:00.

The modeling of energy demand profiles is technically known as quite complex due to the dynamics of the energy flow in buildings. Peak load calculations are generally performed with some assumptions. Therefore, high-capacity energy devices are set up in the buildings, and a higher grid power supply is required. Additionally, the new devices consuming energy have been included in the buildings day by day. Consequently, ANN modeling can help to estimate the current and future energy demand profiles in the buildings.

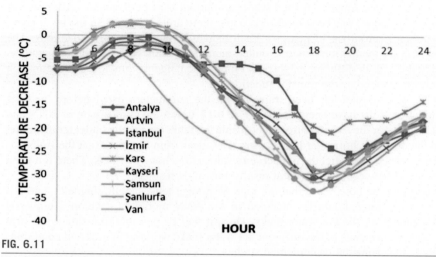

FIG. 6.11

The ANN temperature decrease prediction for the nine cities from Turkey. *ANN*, artificial neural network.

Adapted from Ref. [11].

6.5 Artificial Neural Network Application in Heat Storage System

In each field of engineering and science, there are various applications of the ANN in heat storage. The design of the heat storage systems is depended on many criteria such as the physical conditions of the facility, climatic conditions, and the types of energy source/demand. It is very hard to combine all parameters that affect the design of the heat transfer to optimize the design of the systems. Here, ANN can help to design heat storage systems by considering the wide range of independent parameters. Here, various ANN applications in heat storage techniques are introduced to the readers.

6.5.1 Case Study 1: Thermal Stratification in a Hot Water Tank

As mentioned in the previous chapter, the hot water tanks are one of the common heat storage techniques, and thermal stratification of the hot water tank is the most dominant criterion for their heat storage performance. To determine the degree of the thermal stratification, the temperature distribution inside the tank is required. The temperature distribution inside the tank is generally obtained by numerical or experimental works. While numerical analyses require high-capacity computers and a long time period, the experimental works permit analysis with a limited parameter. The temperature distribution inside the tank and the degree of the thermal stratification can be obtained by the ANN. Erdemir [12] has developed an ANN model to estimate the temperature changes with time inside the tank. The input data for the ANN have been obtained by experimental works in Refs. [13,14].

In this case study, an ANN model that can be an alternative to time-dependent numerical analysis and experimental works in terms of time and results has been created. Thanks to the ANN created when the inlet temperatures and time are given as the input data to the model, the temperature distribution inside the tank and outlet temperatures can be obtained by the ANN.

Feed-forward and backpropagation algorithm has been used as the training algorithm. The least squares method has been chosen as a performance criterion. An ANN of 13 layers has been modeled. While there are 13 neurons in the input layer, the output neuron has 12 neurons. Each hidden layer consists of 13 neurons. The inlet data are mantle inlet, mantle outlet, main inlet, main outlet, eight temperatures inside the tank in different heights, and time. The tangent sigmoid transfer function has been selected as the transfer function.

An ANN model has been trained with the parameters listed before, and the following results have been reached. Fig. 6.12 shows the comparison of ANN and test results, for the temperature values inside the tank, the mantle outlet temperature, and the domestic hot water temperature. It is clear from Fig. 6.12 that the ANN has been achieved to predict the temperature values for a hot water tank. There is a good agreement between experimental results and ANN results.

It should be noted here that the type of storage tank, height, diameter, and climatic conditions should be included in the ANN model to develop a universal ANN model for the hot water tanks. Thus, it will be possible to estimate the heat storage performance of the different hot water tanks working in different conditions without experimental or numerical studies.

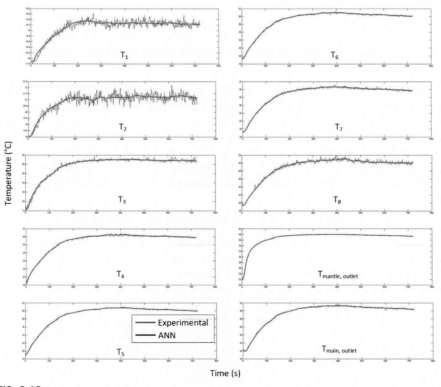

FIG. 6.12

The predicted temperature values from the mantled hot water tank by the ANN. *ANN*, artificial neural network.

6.5.2 Case Study 2: Thermally Stratified Chiller Water Heat Storage Tank

The aim of this case study is to investigate the use of the ANN for thermally stratified heat storage system and borehole heat exchanger for ground source heat pump, based on Ref. [15]. Fig. 6.13 shows the schematics of the studied heat storage system. During the charging period, the inlet water temperature (T_C) from the heat source is 4°C. The return water temperature (T_R) of TES from the load side is 14 °C in the discharging period. The data sets for the ANN have been obtained by the actual operating data of a building located in Tokyo. There is a thermally stratified chilled water heat storage unit that has 2800 m³ of volume. The data acquisition was performed for 1 year at an interval of every 10 min. The measurement points are a total of 20 locations in the vertical direction inside the heat storage tank. Totally, 51,913 data have been gathered.

The ANN has been created via MATLAB R2017a software. The Levenberg–Marquardt algorithm is the training algorithm. The data set has been divided randomly as 70% for training, 15% for test, and 15% for validation. The ANN has been modeled to estimate the temperature gradient inside the tank. The ANN has four layers: input, two hidden, and output layers. Hidden layers have 20 neurons.

The ANN developed has been used for obtaining the temperature gradient inside the tank. The obtained temperature gradient has been used for the borehole heat exchanger. It is examined how to compile a training data set over two case studies of the heat storage system and eight case studies of the borehole heat exchanger system. Table 6.2 gives the results of average value of five training runs for all 10 cases

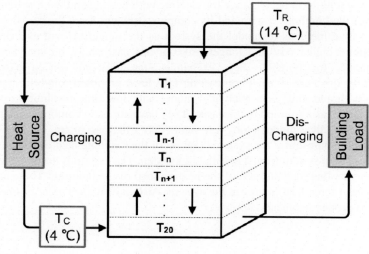

FIG. 6.13

The schematic of the studied system.

Adapted from Ref. [15].

Table 6.2 R^2 and Mean Square Error (MSE) Values for the Studied 10 Cases.

Type	R^2	MSE
Case 1	0.9978	0.0005
Case 2	0.9976	0.0325
Case 3	0.9953	0.0097
Case 4	0.9877	0.0254
Case 5	0.9870	0.0267
Case 6	0.9931	0.0140
Case 7	0.9942	0.0118
Case 8	0.9945	0.0114
Case 9	0.9645	0.0717
Case 10	0.9907	0.0198

Modified from Ref. [15].

studied. It is clear from Table 6.2 that the ANN modeling has been performed precisely. The obtained temperature distribution data from ANN have provided sufficient input data for the further studies.

6.5.3 Case Study 3: Analytical—Artificial Neural Network Hybrid Model for Borehole Heat Exchanger

The borehole heat storage systems are one of the significant examples of high-capacity and long-term heat storage for meeting both heating and cooling demands. The size of the borehole system and its complex structure force the researchers to find alternative methods to model the system as the numerical and experimental works require higher cost and longer time. In this case study, an analytical and hybrid model for borehole heat exchanger has been introduced based on Ref. [16]. In this case study, Puttige et al. [16] have proposed a hybrid approach that uses a simple long-term analytical model with low time resolution to guide the ANN model as a solution to this issue. They have aimed to combine the capabilities of the analytical model's long-term and ANN's high accuracy.

Fig. 6.14 demonstrates the system diagram of the borehole heat exchanger integrated heat pump system from a hospital in Sweden. The system is designed to meet 95% of the cooling load and 5% of the heating load of the building. There are 125 boreholes and three heat pumps in the system. Also, a heat exchanger has been integrated to produce domestic hot water.

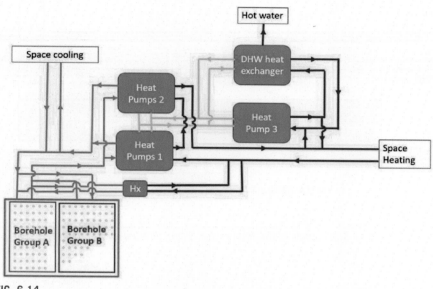

FIG. 6.14

The schematic view of the borehole heat—exchanged integrated heat pump system.

Adapted from Ref. [16].

Fig. 6.15 illustrates the input and outputs of the ANN model on a timeline. There are 37 inputs and 2 outputs which are heat loads at the current time step for two groups of boreholes. The number of hidden layers is one that has 27 neurons. While the activation function in the hidden layer is the hyperbolic tangent, the linear activation function has been used in the output layer. The ANN has been created by using MATLAB 2020.

The RMSE values of the ANN model for training, testing, and validation are 12, 19.3, and 26.6 kW, which correspond to 3.2%, 6.1%, and 7.3%, respectively. Fig. 6.16 shows the comparison of the results for ANN and two analytical models. As seen from Fig. 6.16, the ANN adaptation into the calculations has reduced the RMSE values significantly. The analytical models and ANN have been modeled by using data for 2017 and 2018. The hybrid model has been validated with 2019 data. All models have been compared for the data of 2020. The relative RMSE values for 2020 for ANA, ANA_calib, and hybrid models are 21.9%, 13.2%, and 6.3%, respectively. The results show that the error of the hybrid model is significantly lower than the analytical models.

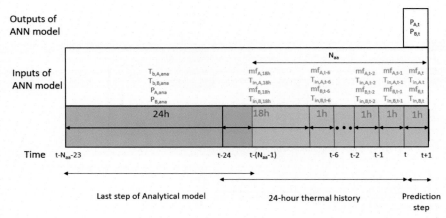

FIG. 6.15

Illustration of input and outputs of the ANN model on the timeline. *ANN*, artificial neural network.

Adapted from Ref. [16].

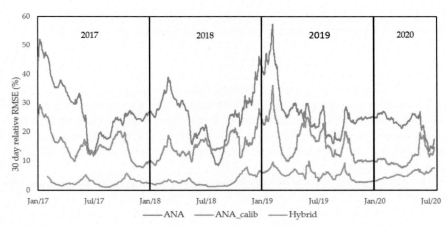

FIG. 6.16

The comparison of the 30-day relative RMSE for two analytical models and ANN. *ANN*, artificial neural network; *RMSE*, root mean square error.

Adapted from Ref. [16].

6.6 Closing Remarks

In this chapter, a significant subject (so-called AI) is considered for energy applications, particularly for heat storage applications. One should remember that AI applications can be integrated into heat storage systems in many ways. For instance, AI can be used for predicting the performance criteria of heat storage systems under different climatic conditions where such conditions directly affect the practical performance of the systems. This chapter provides an introduction to AI methods and their role in heat storage systems for various purposes, ranging from system analysis to performance assessment. More importantly, some case studies are also presented to illustrate how effectively AI methods (including ANN approaches) are to be employed for heat storage systems and their applications.

References

[1] I. Dincer, C. Acar, A review on clean energy solutions for better sustainability, Int. J. Energy Res. 39 (2015) 585–606, https://doi.org/10.1002/er.3329.

[2] K. Abhishek, M.P. Singh, S. Ghosh, A. Anand, Weather forecasting model using artificial neural network, Procedia Technol. 4 (2012) 311–318, https://doi.org/10.1016/j.protcy.2012.05.047.

[3] M. Temizyurek, F. Dadaser-Celik, Modelling the effects of meteorological parameters on water temperature using artificial neural networks, Water Sci. Technol. 77 (2018) 1724–1733, https://doi.org/10.2166/wst.2018.058.

[4] P. Rajendra, K.V.N. Murthy, A. Subbarao, R. Boadh, Use of ANN models in the prediction of meteorological data, Model. Earth Syst. Environ. 5 (2019) 1051–1058, https://doi.org/10.1007/s40808-019-00590-2.

[5] H.K. Ghritlahre, P. Chandrakar, A. Ahmad, A comprehensive review on performance prediction of solar air heaters using artificial neural network, Ann. Data Sci. (2019), https://doi.org/10.1007/s40745-019-00236-1.

[6] S.L. Muhammad, W.A.W.Z. Abidin, W.Y. Chai, A. Baharun, T. Masri, Development of wind mapping based on artificial neural network (ANN) for energy exploration in Sarawak, Int. J. Renew. Energy Resour. 4 (2014) 618–627.

[7] A. Keçebaş, İ. Yabanova, M. Yumurtacı, Artificial neural network modeling of geothermal district heating system thought exergy analysis, Energy Convers. Manag. 64 (2012) 206–212, https://doi.org/10.1016/j.enconman.2012.06.002.

[8] G.B. GridWatch, Electricity National Grid Demand and Output, 2020. https://gridwatch.co.uk/. (Accessed 4 January 2020).

[9] M.N.Q. Macedo, J.J.M. Galo, L.A.L. de Almeida, A.C. de Lima, Demand side management using artificial neural networks in a smart grid environment, Renew. Sustain. Energy Rev. 41 (2015) 128–133, https://doi.org/10.1016/j.rser.2014.08.035.

[10] M. Yalcintas, S. Akkurt, Artificial neural networks applications in building energy predictions and a case study for tropical climates, Int. J. Energy Res. 29 (2005) 891–901, https://doi.org/10.1002/er.1105.

[11] D. Erdemir, T. Ayata, Prediction of temperature decreasing on a green roof by using artificial neural network, Appl. Therm. Eng. 112 (2017), https://doi.org/10.1016/j.applthermaleng.2016.10.145.

[12] D. Erdemir, Experimental Study of Thermal Stratification on Mantled Hot Water Tank, Evaluation in the Terms of Second Law's Thermodynamics and Ann Modelling, Erciyes University, 2013.

[13] D. Erdemir, N. Altuntop, Improved thermal stratification with obstacles placed inside the vertical mantled hot water tanks, Appl. Therm. Eng. 100 (2016), https://doi.org/10.1016/j.applthermaleng.2016.01.069.

[14] D. Erdemir, N. Altuntop, Effect of thermal stratification on energy and exergy in vertical mantled heat exchanger, Int. J. Exergy 20 (2016), https://doi.org/10.1504/IJEX.2016.076681.

[15] D. Lee, R. Ooka, S. Ikeda, W. Choi, Case study of ANN modeling of stratified thermal energy storage and ground source heat pump systems for model predictive control, in: 4th Int. Conf. Build. Energy, Environ., 2018. Melborn, Australia.

[16] A.R. Puttige, S. Andersson, R. Östin, T. Olofsson, A novel analytical-ANN hybrid model for borehole heat exchanger, Energies 13 (2020), https://doi.org/10.3390/en13236213.

Conclusions and Future Directions

7

7.1 Conclusions

In this book, heat storage systems for building applications are studied and discussed in depth, considering various aspects of their development, analyses, modeling, evaluation, and implementation. Firstly, the fundamentals and key concepts of energy storage systems are introduced. The advantages of the energy storage system are discussed with practical applications. The answer to the question of why energy storage systems are needed is explained with detailed examples from various sectors. The energy storage methods are introduced briefly. In the second chapter, the heat storage methods are elucidated in depth with examples of practical applications considering their advantages and disadvantages. Where and how each heat storage method is able to be used is explained. Their potentials to manage energy supply and demand profiles and reducing energy costs are discussed. In the third chapter, the energy systems used in the buildings are introduced to the readers. Based on the energy consumptions of these systems, the need for heat storage in buildings is emphasized. In the fourth chapter, the methods of system analysis, modeling, and simulation for the heat storage systems are explained with illustrative examples from the practical applications of heat storage in the buildings. The fundamentals of thermodynamics, heat transfer, and numerical simulation for the heat storage systems are explained. In the fifth chapter, the case studies for the heat storage system used in the buildings have been presented. The systems in the case studies are evaluated through the background introduced in the fourth chapter. Lastly, in the sixth chapter, the artificial neural network (ANN) application in heat storage systems has been introduced by emphasizing the need for ANN modeling in heat storage systems.

Energy management in the building and communities is recognized as one of the critical topics for better sustainability since the buildings are responsible for one-third of total energy consumption in the world. In the past, the energy management in the buildings was not complicated to model and control, as generally conventional fossil fuel—driven systems and limited renewable energy systems are used. However, today, energy management in buildings is quite complicated to model and control. There are many reasons for complex energy management, as listed in the following:

- The integration of renewables in the building energy system has been increasing day by day. In the past, renewables were used for meeting specific needs, such as solar energy systems for hot water and heating, and wind for electricity. Today, they are using for meeting many demand items thanks to multigeneration systems. Also, a few renewable energy sources can be used at the same time.
- Adding new energy-consuming items is one of the problems facing energy management processes. Every day, a new device, machine, and processes have been included in our daily lives, such as electric vehicles, fresh air supply systems, and so on. In the near future, the energy demand for electric vehicles will be a significant problem for almost all existing buildings. As another example, with the COVID-19 pandemic, the fresh air supply for the closed spaces in the buildings brings a tremendous energy demand. Such additional energy demand items have been causing more complicated energy management in the building and communities.
- The need for an increase in the capacities of the devices used in the buildings is another problem in energy management. Due to changes in internal and external dynamics of the buildings, the capacity of the devices is needed to be increased. For example, the capacity of air-conditioning (AC) systems, which is one of the most energy-consuming items in the buildings, is needed to be increased due to increasing outdoor temperature and the increase of time people spend indoors. Such dynamics make the energy management models more complicated.
- One of the most straightforward applications that can be done for efficient energy use is to use devices with higher efficiency in buildings, such as LED lighting equipment, chillers that have higher COP, devices that have higher energy efficiency indexes (A, A+, A++, etc.), and heat insulations. The potentials of these systems and devices in reducing energy consumption and cost should be included in the effective energy management models.

Recovering the energy losses is one of the essential duties for energy management in buildings. However, the determination of the potential of energy recovery and its applications will bring a vastly complex structure for energy management models.

Renewable energy sources are included in building energy systems for reducing carbon emission and hence greenhouse gas emissions. The integration of energy storage is essential to benefit from the renewable energy source in a more effective way. To extend the duration of the availability of renewables and balance the fluctuating behavior of the availability, energy storage and heat storage systems are the unique solution. Especially in the buildings, heat storage systems are more convenient as heating and cooling are the leading demand items.

Energy storage systems and heat storage systems, especially for buildings, offer a significant potential to help solve these issues defined earlier. Heat storage systems can solve the issues in capacity increases in heating and cooling systems without requiring high-capacity devices. For example, due to the COVID-19, there is an urgent need for capacity increases for fresh air supply systems. This brings a huge capacity increasing and energy consumption. Heat storage systems can meet this additional energy capacity increasing by using existing heating and cooling system.

The cooling demands of the buildings are ever increasing due to increasing outdoor temperatures and the increase of time people spend indoors. Also, to meet the increase in the cooling loads, additional devices are required. The additional devices require considerable energy consumption and power supply from the grid. The cold heat storage systems such as ice storage systems can solve the increase in capacity problem without requiring any additional device or power supply from the grid. The heat can be charged into cold storage medium during the off-peak period, and the stored energy can be used for meeting additional cooling demands.

As heat storage systems can change the operating period of the devices, they can increase the efficiency of the systems. For example, cold heat storage systems can shift the cooling loads from daytime to nighttime. The COP of the chillers increases due to changes in the environmental temperature between daytime and nighttime. Thus, the energy consumption reduces with the increasing COP.

Heat recovery is an essential need for the buildings, like in recycling of the wastes. However, the recovered energy may not be used or needed when it is recovered. The type of energy recovered is generally the heat. Therefore, heat storage systems can increase the effectiveness of energy recovery.

As seen from these examples given before, heat storage systems have a significant potential to manage the energy demand profiles in buildings. However, in many energy management algorithms, systems, and processes, the energy storage systems are not included. When the potential of the heat storage systems is considered in the building energy supply and demand profiles, they should be included in the energy management algorithm, system, and processes. Thus, energy management can be performed more effectively.

One of the primary purposes of heat storage systems is to reduce the dependency on the grid power supply. They can achieve this in many ways. For example, since heat storage systems can extend the availability of renewable energy sources, they reduce the grid energy demands and costs. As another example, the heat storage system can reduce energy consumption costs by shifting the heating and cooling load from the peak periods to off-peak periods. The potential of the heat storage systems in reducing energy costs is crystal clear to everyone. Today, they are proved technologies. The biggest problem in using heat storage systems is finding an available space for the heat storage units. During the design phase of the buildings, the required space can be created easily by considering the statics and dynamics of the building. Therefore, the heat storage systems can be included in the building energy systems with the standards. Thus, the use of the heat storage systems can be extended. On the other hand, it is hard to create a space for the heat storage units in the existing buildings. However, while the underground storage tanks can be used for high-capacity storage, the lower-volume indoor storage tanks can be placed in the building's basement. Partial heat storage strategies can be useful for the existing buildings if there is no enough space for full storage.

In addition to reducing energy costs, heat storage systems have unique advantages for reducing the capacity of the heating, ventilation, and air-conditioning (HVAC) systems. Load leveling strategy is the most convenient strategy to minimize the capacity of the HVAC systems. With the load leveling strategy, the energy management of the buildings is performed efficiently, as the energy demand is constant. The decreasing device capacities also reduce the size, cost, and capacity of the auxiliary devices used in the buildings, such as generators, cable sizes, grid power supply, and so on. Therefore, the advantages of the heat storage systems should be assessed from a broad perspective.

In the certain buildings where cooling is significant and critical, such as data processing centers, supercomputer centers, stem cell and embryo centers, museums, and so on, there are more than one backup chillers and power generators used for the chillers. These additional backup devices bring along with it a substantial initial investment and service costs. Ice storage or other cold heat storage techniques can be used for emergencies. Thus, both cooling costs and the costs of devices used for emergencies can be reduced. It is important to note that the heat storage methods used for the different purposes listed before during normal periods can be used to meet the extreme burden that occurs in emergency situations. During the COVID-19 pandemic, the capacities of hospitals have been exceeded substantially around the world. However, the capacities of AC systems in hospitals are not enough to meet this unexpected demand. While cold heat storage systems can be used for meeting the unexpected high cooling demands during emergencies, they can reduce the cooling costs and shift the peak loads to off-peak hours during the normal days.

7.2 Future Directions

Heat storage methods are, by the way, not new but have been used by people since ancient times. Probably, heat storage is as old as civilization itself. From the past to present, energy has been an important need for people in any form of it such as thermal, mechanic, electricity, and so on. Since recorded times, people have harvested ice and store it for later use for cooling purposes and have collected woods for the later heating purposes. As depicted in Fig. 7.1, the heat storage systems have shown significant developments, parallel to developing technology and increasing energy consumptions. The energy consumption in the world has increased substantially with the industrial revolution. Today, energy is an essential need for people's life and takes part in everything from mobile devices to vehicles, from thermal comfort devices to space technologies. Increased and changing energy use forced people to use energy efficiently. Therewithal, the milestones in the energy policies and crises have promoted the use of renewables and energy storage systems. For example, the oil crisis of the early 1970s has forced people to use renewable and energy storage systems, such as solar domestic hot water systems. Heat storage in the rock bed and solar ponds has started to use in practical applications for heating purposes.

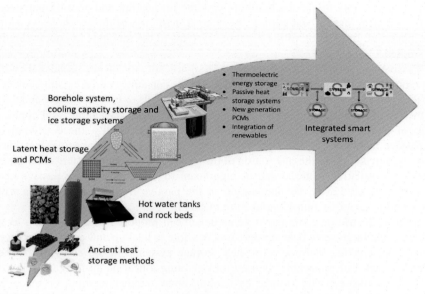

Borehole system,
cooling capacity storage and
ice storage systems

Thermoelectric
energy storage
• Passive heat
storage systems
• New generation
PCMs
• Integration of
renewables

Integrated smart
systems

Latent heat storage
and PCMs

Hot water tanks
and rock beds

Ancient heat
storage methods

FIG. 7.1

Evaluation of the heat storage systems.

Generally, the common materials in nature were being used as storage materials or mediums, such as water, rock, sand, and so on. Higher-capacity heat storage systems and materials have been required with the increase of heat storage applications. The developments and practical applications of latent heat storage systems and phase change materials have increased. Today, many phase change materials have been used in practical applications, and also more are under development.

Water/ice is one the most effective phase change materials for cold capacity storage due to the high latent heat of the phase change of water/ice. The use of ACs, and also energy consumption due to them, has been substantially increasing with the increasing thermal comfort conditions at the global scale. To reduce both energy costs and consumption of the ACs, ice storage and other cold capacity storage techniques are a unique solution. Their integration into the building HVAC systems is ever increasing. To design an effective ice storage system, the geometry of the ice capsules, also called ice balls, should be studied for the encapsulated systems. Also, modular and thermally enhanced storage tanks are wanted by the sector. Although the ice slurry systems provide better performance than other ice storage techniques, there are limited applications of ice slurry systems in terms of practical relevance. Breakthrough developments are needed in the ice slurry systems for extending their applications.

High-capacity and long-term heat storage needs have forced the engineers to develop the alternative heat storage techniques. Borehole heat storage systems are one of the most common heat storage techniques for both high capacity and long term. It is possible to use the borehole systems for heating and cooling purposes. With increasing use of heat pumps, the interest in the use of the boreholes systems has increased. Integration of the borehole systems to the integrated systems should be studied.

Passive heat storage techniques provide significant advantages because they do not require any additional device or increase in the capacities of the existing devices for the buildings. Embedded storage materials into walls or construction foundation of the buildings are applying the buildings. Development, packaging, and applications of the storage mediums should be researched. Especially, the development of passive heat storage systems for both heating and cooling capacities becomes more significant. Also, the integration of the passive heat storage systems into green roofs and insulation material can be significant for the future of the passive heat storage systems. Lastly, the integration of the passive heat storage systems with renewables is another future application of them.

Thermoelectric heat storage systems have been proposed as a promising technique for the large-scale heat storage. It is possible to heat up and cool down power generation with the thermoelectric heat storage systems. In the charging period, electricity (grid or renewable) is used for operating a heat pump for generation of heat for heating capacity storage by cooling down another medium or space. In the discharging period, the temperature difference between hot and cold mediums or spaces is used for running a heat engine. Thus, the electricity is generated. Due to its unique advantages of thermoelectric heat storage systems, in the future, they can be a key option for managing the building energy demands.

The integration of the renewable energy systems into heat storage systems, like in almost all energy systems, is a significant task for making the systems GREEN. Especially, for the new generation for heat storage systems, which are passive heat storage systems and thermoelectric heat storage systems, the integration of the renewables is essential. On the other hand, the development of the novel heat storage materials is a critical topic for the future of these systems.

We are living the integration era because integration is essential for almost everything to increase efficiency and productivity. That is why, things are now more integrated. Today, one of the most common integration terms is the artificial intelligence (AI). We are talking about smart version of all systems, devices, and processes.

Today, we are talking about zero carbon emission systems, and the only way to achieve this issue is the renewable energy—integrated multigeneration energy systems. Energy storage systems are essential part of the integrated systems, as they can make the integrated systems more manageable. The useful outputs of most of the integrated systems consist of heating and cooling of the buildings or processes. Therefore, heat storage systems are the significant option for the integrated systems. They can help to reduce the capacity and size of the system elements. Renewable energy sources are the heart of the integrated systems on the way of zero carbon. To regulate the fluctuating behavior of the renewables, the integration of the energy storage techniques is an only solution. They also extend to available time of the renewables when they are not active. Consequently, the energy storage methods should be included in the integrated energy systems for better management, lower cost, higher efficiency, and better renewable integration.

Today, AI is recognized as a potential area of science and technology and expected to play an essential part of our lives. The future of people will more depend on AI technologies and their immediate implementations. AI applications can be integrated into heat storage systems in many ways. For example, AI can be used for determining the performance of heat storage systems in different climatic conditions, as the climatic conditions directly affect the performance of the heat storage systems. Many AI models have been introduced for determining the heating and cooling load of buildings in different climatic conditions. Also, AI can be used for predicting future cases in the long and short terms. There are numerous AI applications regarding how meteorological data will change in the future. On the other hand, it is also used for managing the subsystems according to the demand or supply profiles in the short term. For example, the use of the stored heat can be extended, or the charging period can be shorted according to current outdoor conditions.

AI can be used in heat storage systems for real-time control of the systems. This application of AI in heat storage systems can be explained through an example. The building heating and cooling system are designed according to the peak load of the building. The operation of the heat storage systems is generally set for specific times. The systems work on and off modes in certain periods. Consider a heat storage system that gets weather forecasts and planning how to use the stored energy. This can occur with the integration of AI. For example, consider such an ice storage system working with the partial storage strategy. The peak cooling load of the buildings is seen in a very limited time, a couple of hours in a day and a couple of weeks in the summer seasons. The design and utilization strategy of the ice storage systems is adjusted according to the peak cooling load of the buildings. According to weather conditions, the stored energy can be used for meeting all the cooling needs of the building. Consequently, an AI-based real-time control mechanism can help to increase the effectiveness of the heat storage systems significantly.

Nomenclature

A	area (m^2)
c	specific heat (kJ/kg K)
D	diameter (m)
e	specific energy (kJ/kg)
E	energy (kJ or kWh)
$\dot{E}$	rate of energy (kW)
ex	specific exergy (kJ/kg)
Ex	exergy (kJ and kWh)
$\dot{Ex}$	rate of exergy (kW)
F	volume fraction of fluid or the view factor for radiation heat transfer
h	enthalpy (kJ/kg) or convective heat transfer coefficient (W/m^2 K)
H	total enthalpy (kJ) or height (m)
k	heat conductivity (W/m K)
K	permeability
KE	kinetic energy (kJ)
L	length (m)
m	mass (kg)
$\dot{m}$	mass flow rate (kg/s)
P	pressure (kPa)
PE	potential energy (kJ)
q	heat transfer per area (kJ/m^2)
$\dot{q}$	heat transfer rate per area (kW/m^2)
Q	heat transfer (kJ or kWh)
$\dot{Q}$	rate of heat transfer (kW)
R	thermal resistance ($kW/m^2 K$)
s	entropy (kJ/kg)
S	total entropy (kJ)
$\dot{S}$	rate of total entropy (kW)
t	time (s or h)
T	temperature (K or °C)
u	internal energy (kJ/kg) or velocity of the working fluid (m/s)
U	total internal energy (kJ)
V	volume (m^3)
$\dot{V}$	volumetric flow rate (m^3/s)
$\dot{W}$	rate of work (kW)
x, X	position, distance (m)

Greek Letters

α	absorptivity
ε	emissivity
η	energy efficiency
μ	dynamic viscosity (Ns/m^2)
ν	specific volume (m^3/kg) or kinematic viscosity (m^2/s)
ρ	density (kg/m^3) or reflectivity
σ	Stefan–Boltzmann constant
τ	transmissivity
ψ	exergy efficiency
φ	porosity

Subscripts

0	reference conditions
ave	average
b	boundary
boiler	boiler
bottom	bottom of the tank
bulk	bulk temperature of fluid
cap	capsule
ch	charging period
comp	compressor
con	condenser
dest	destruction
disch	discharging period
e	electricity or equivalent
env	environment
eva	evaporator
ex	exergy
f	fluid
final	final
flux	heat flux
gain	gain
gen	generation
HTF	heat transfer fluid
hws	hot water storage tank
ice	phase of ice
in, inlet	inlet
initial, init	initial
is	isentropic
loss	loss
m	fully mixed
mantle	mantle
net	net
out	outlet
ove	overall
p	constant pressure

pcm	phase change material
pump	pump
R	refrigeration
s	source or solid
sb	solid block
sen	sensible
sf	solid/fluid phase changing
sh	shaft
sm	storage medium
st	storing period
sur	surface
sys	system
$t=0$	initial conditions of any period
tank	storage tank
top	top of the tank
tur, turbine	turbine
valve	valve
void	volume of void in the porous medium

Abbreviations

AI	artificial intelligence
ANN	artificial neural network
CFD	computational fluid mechanics
EBE	energy balance equation
EnBE	entropy balance equation
ES	energy storage
ExBE	exergy balance equation
FLT	first law of thermodynamics
HS	heat storage
MBE	mass balance equation
PCM	phase change material
SDHWS	solar domestic hot water system
SLT	second law of thermodynamics
TLT	third law of thermodynamics

Index

'Note: Page numbers followed by "f" indicate figures and those followed by "t" indicate tables.'

Printed in the United States
by Baker & Taylor Publisher Services